CLEANUP

CLEANUP

How Repentance Restores Relationships

STEVE SHORES

WIPF & STOCK · Eugene, Oregon

CLEANUP
How Repentance Restores Relationships

Copyright © 2022 Steve Shores. All rights reserved. Except for brief quotations in critical publications or reviews, no part of this book may be reproduced in any manner without prior written permission from the publisher. Write: Permissions, Wipf and Stock Publishers, 199 W. 8th Ave., Suite 3, Eugene, OR 97401.

Wipf & Stock
An Imprint of Wipf and Stock Publishers
199 W. 8th Ave., Suite 3
Eugene, OR 97401

www.wipfandstock.com

PAPERBACK ISBN: 978-1-6667-3456-0
HARDCOVER ISBN: 978-1-6667-9056-6
EBOOK ISBN: 978-1-6667-9057-3

03/30/22

Unless otherwise noted, Scripture quotations taken from the (NASB®) New American Standard Bible®, Copyright © 1960, 1971, 1977, 1995, 2020 by The Lockman Foundation. Used by permission. All rights reserved. www.lockman.org.

Citations marked ESV® are from The ESV Bible (The Holy Bible, English Standard Version®), copyright © 2001 by Crossway, a publishing ministry of Good News Publishers. Used by permission. All rights reserved.

Excerpt from "Burnt Norton" from *Four Quartets* by T. S. Eliot. Copyright © 1952 by Houghton Mifflin Harcourt Publishing Company, renewed 1980 by Esme Valerie Eliot. Reprinted by permission of Mariner Books, an imprint of HarperCollins Publishers LLC. All rights reserved.

All stories from my counseling practice have been modified to protect privacy and confidentiality.

Dedicated to
Katy Van Wyk
Jenny Pfost
Christy Shores

For the mind set on the flesh is death, but the mind set on the Spirit is life and peace. (Rom 8:6)

Contents

Preface		ix
Acknowledgments		xi
Chapter 1	Closed People and Relational Debt	1
Chapter 2	Pain (Fallen World 1)	17
Chapter 3	Fear and Shame (Fallen World 2)	31
Chapter 4	Flesh (Anesthesia)	42
Chapter 5	Control and Safety (Sleep)	48
Chapter 6	The Role of Repentance (Waking)	57
Chapter 7	Love (Reaching)	66
Chapter 8	Synthesis	81
Bibliography		93

Preface

For a fuller explanation of the role of this volume, consult the preface of volume 1, *Stuck: How We Are Reverse Born Again*. To get you oriented to the immediate book in hand, I refer you back to the epigraph and its sober yet hopeful picture of a penetrating tension between the flesh within us and the Holy Spirit, who is still deeper inside and more fundamental to who Christians actually are. When we add the reality that "the flesh sets its desire against the Spirit, and the Spirit against the flesh; for these are in opposition to one another" (Gal 5:17), we get a glimpse of the intensity of an internal battle that teeters between "death" on the one side and "life and peace" on the other. These are not small stakes, and their effects emerge, as we will see, most dramatically in the everyday theater of human relationships. In other words, how we fare in this battle determines the quality of love (or lack of it) in our connections with one another.

Acknowledgments

For twenty-eight years, I have had the immense privilege of meeting weekly with a group of men for open discussion about anything at all that we might find challenging about living as Christians who are far from home. Robert Broome, Scott Broome, Andrew Kaiser, and Dave Shores have, each in his own Spirit-led way, kept me from walking off many ledges in my thinking life and my emotional life. Their love and wisdom scintillate through these pages.

I would also like to thank Kara Barlow, my copy editor. I'm grateful that she has consistently provided a wise and clarifying perspective. Not to mention an eye for detail that I lack. These books would be much the poorer without her.

CHAPTER 1

Closed People and Relational Debt

Some years ago, I traveled to St. John in the US Virgin Islands. At a local restaurant, I listened while the owner passionately warned of the environmental impact of sunscreen: "What do you think?! It's a petroleum-based product. Tourists slather it on and swim all over the place. All those petrochemicals go right into the ecosystem and bleach the coral." That night, I went to a lecture on the ecological threat posed by the lionfish, an invasive species probably brought from the Indian Ocean in ships' bilgewater. Lionfish have no natural enemies in the Caribbean, are serious eaters, and will devour anything up to three-fourths their own size. They're basically voracious vacuum cleaners. I continued to learn: later in the week, a local activist told me that developers on St. John often don't follow requirements to install barriers for diverting runoff from newly paved driveways and parking lots. The unimpeded runoff finds its way into the ocean and into the coral, with devastating effects.

Let's take that last example and work with it. In the absence of diverting barriers, runoff carries motor oil and fertilizers into the waters around the island. This noxious soup overburdens the aquatic world with substances it can't neutralize. The result is the nasty reality called coral bleaching referred to by the restaurant owner. Bleaching occurs because the coastal water's new chemical

CLEANUP

toxins distress critters called zooxanthellae and run them out of the coral. A vital symbiotic relationship ends, and the coral begins to die. Dead coral is no longer a viable ecosystem, so fish, turtles, lobsters, crabs, worms, octopi, cuttlefish, and other animals and plants lose vital habitat in a devastating ripple effect.

We'll now work with our ecological example some more, using it to deepen our understanding of the internal world of people. Within every human, a set of routines called "the flesh" exerts itself. These habitual routines form strategic patterns of behavior aimed at self-protection. The result is that the "good life" is now defined as follows: "When I feel comfortable and safe on terms that make sense to me, that's the good life." The problem with that reasoning is that one's sense-making capacities may be deeply, subtly influenced by the flesh. With that in mind, let's imagine that human flesh patterns can be viewed from an ecological standpoint. Just as the physical environment needs a balance of resources to remain healthy, the same is true of human relationships. But the flesh is not thinking relationally; it's thinking selfishly. Its self-protective patterns "pave over" acres of relational territory just as monied self-interest paves over the landscape of St. John to make "development" easier.

In an ideal world, the honesty and vulnerability of flesh-free relationships would keep the relational ecology fresh between human beings. But the flesh is quick to frown down the honest and the vulnerable, because these invite us to drop our defenses. Honesty and vulnerability crash against the selfishness of the fallen heart. The flesh flattens ("paves") relationships into arrangements (see introduction to the first volume, *Stuck*) of undiscussed procedures and maneuvers based on domination, fear, hiding, manipulation, defense, neutralization. In other words, the flesh develops predictable habits that don't allow for honest, potentially creative risks in relationships.

We'll define the flesh as understood biblically later on; for now, it's enough to know that the flesh is our inborn propensity to sin, and it is well described as our allergy to God. The idea is that of a deep recoil from God inside the "me" who wants to skipper my own ship.

CLOSED PEOPLE AND RELATIONAL DEBT

The flesh is defensive: "Briefly put, it is the creature who, turned from its Lord, orbits about itself and intends to make its own way."[1] Under the illusion of having dethroned God, it seeks that personal safety and control we've been discussing. It exclusively asks, "Am I getting the outcomes I deem will make me feel safe and in control on my terms?" This limited, me-first focus flattens relationships into mere arrangements along the lines of "I do my thing, and you're free to do yours as long as you don't step on my outcomes." Relationships cannot be mutual in any real way under such conditions. For example, if I'm always shepherding my outcomes along and keeping danger away from my "flock," I won't give you the full attention that a real relationship needs. The urge for safety and control creates a situation analogous to that environmental runoff we've described. My pursuing my own outcomes always causes a bad relational impact. As that impact goes undiscussed and unresolved, these suppressed forces "flatten" arrangements between people. These arrangements pose as relationships while their toxins seep into the relational "ecosystem."

For example, let's imagine a husband who's prone to bouts of intense anger. The eruptions are meant to let others know that he's not to be trifled with. Deploying these strategies helps him feel he's in control. But he's oblivious to their impact on his wife. When she tries to be honest with him about her pain and fear, he reads her feedback as an intense threat to his campaign for safety and control (a campaign to which he's blind), so he blows up. When she points out, "This is the scary behavior I'm talking about," he accuses her of thinking she's perfect. Losing her battle with herself, she deploys one of her own strategies: sarcasm. "Oh, so I guess you think your little tantrum is the pinnacle of Christ-likeness!" To which he yells, "Oh, so I guess the other day when *you* were angry, that was perfectly justified!" And so on.

It's not hard to predict how the argument will continue. And it's not hard to predict that toxic thoughts and emotions toward one another will dominate this marriage for the next hours or days. The tragedy and toxicity seep into their hearts and darken their very

---

1. Käsemann, *On Being a Disciple*, 98.

CLEANUP

prayers. Such relational "runoff" can be thought of as a rising debt made up of bypassed wounds. These wounds of the heart are of two kinds: wounds we've received and wounds we've perpetrated. Perpetration introduces the idea of an uneasy conscience, which becomes part of the load under which the relationship groans. The "chemistry" of the relationship gradually erodes as it takes on more and more debt. By "debt," I mean that the longer a person is self-protective in a relationship, the more he or she "owes" the relationship a compensating, corrective "payment." If this payment is made early (through reflection, risk, repentance, growth, godly sorrow, behavioral change), the relationship has a chance to recover. But if the debt mounts for years and years, the structure of the relationship is less and less able to support the debt. Just as the coral reef eventually cannot work off the load of pollution if it continues unchecked, so a relationship becomes less and less able to counteract the toxic buildup that flesh patterns impose on it.

A similar analogy is that of a car that keeps running for years without an oil change. The lack of fresh oil slowly builds up a debt to the engine. That structure comes under the stress of a larger and larger "unpaid bill" consisting of the clean oil it regularly needs to maintain its integrity. Eventually, the structural debt will be paid in the form of engine inefficiency and eventual failure. Relationships have a structure, too. They're meant to be built on real love, respect, honesty, compassion, risk-taking, vulnerability, prayer. The relational structure loses strength and suppleness when pride, unkindness, selfishness, distance, manipulation, lying, power games, etc. result in "unpaid bills"—i.e., missed opportunities for relational nourishment.

Dead coral is lifeless and pale, an underwater firmament of skull-colored tragedy, a poisoned badlands, a desert of sorrow (bones without hope, as in Ezek 37:11). Just so, relationships too easily become empty, thin, hollowed out. They devolve into sad, leaning frameworks, tottering dead zones blighted from within. Coral dies owing to lack of ecological knowledge. The same is true of relationships, which also, as we're seeing, have an ecology. But coming to know this, coming to *see*, presents a challenge. Seeing clearly disturbs the hidden procedures we use to survive, bringing to

CLOSED PEOPLE AND RELATIONAL DEBT

light their relational costs. Something in the human does not want to know. Most don't want to see the creeping deadness in their relationships. Those who do see tend to be lonely, because their vantage point threatens those who prefer the convenience of blindness. Aldo Leopold, one of our early writers on ecology, puts it this way:

> One of the penalties of an ecological education is that one lives alone in a world of wounds. Much of the damage inflicted on land is quite invisible to the layman. An ecologist must either harden his shell and make believe that the consequences of science are none of his business, or he must be the doctor who sees the marks of death in a community that believes itself well, and does not want to be told otherwise. One sometimes envies the ignorance of those who rhapsodize about the lovely countryside in process of losing its topsoil, or afflicted with some degenerative disease in its water systems, fauna, or flora.[2]

Just as the physical ecologist may suffer loneliness as the cost of seeing damage others find hard to spot, so the "relational ecologist" can feel alienated from his or her fellows. The price of saying what one sees "can cause deep and permanent damage to the most intimate of relationships, as Jesus forewarned" in Luke 12:51–53.[3] The questions become difficult: Are you in a relational dead zone? Do you contribute deadness of your own? Because the flesh is always seeking to express itself, the answers are almost inevitably yes and yes. Where do we get the courage to face these questions?

By way of seeking an answer, let's again take the questions into a specific type of relationship: marriage. The couple we just saw tragically locking horns is not rare. Many couples go for years accruing a relational debt until it becomes large enough to suffocate the marriage. One way to understand the prevalence of divorce is to use this idea of relational debt. Divorce happens when the relationship can no longer tolerate what one or both partners owe to the structure of the relationship. The emotional "payments" become too great, and the relationship fractures.

---

2. Leopold, *A Sand County Almanac*, cited in Wirzba, *Paradise of God*, 97.
3. Willard, *Renovation*, 98.

CLEANUP

Take, for example, a husband who drives a truck for a living. Most of his trips are long-haul, so he is gone three-fourths of the time. When at home, he's so stressed out from fighting traffic and deadlines that he retreats into himself, spending hours nestled in his recliner, cocooned with the television. It's as though he's still in the cab of the truck, isolated and protected. His wife, longing for emotional connection, brings up, from time to time, the distance between them. He gets defensive, saying things like "I can't do anything right. I could be some super-sensitive metrosexual, and it still wouldn't be enough for you!" His wife then shuts down and turns her emotional needs toward parenting their daughter. The husband, in turn, feels left out of the increasing bond between his wife and daughter. He hardens his heart so it won't hurt. He finds reasons to stay on the road longer, which increases his wife's isolation, which tempts her to draw even closer to their daughter. The relational debt in the marriage is now skyrocketing, yet the defensive patterns continue.

Why do so many couples carry such a risky load of relational debt? One answer is that the toxicity can be camouflaged and denied through a haze of self-deception. As Leopold said a couple paragraphs ago, "One sometimes envies the ignorance of those who rhapsodize about the lovely countryside in process of losing its topsoil."[4] In other words, the truth is bypassed, and the fact that it is bypassed is also bypassed.[5] This bypassing is simply part of how the flesh works. The flesh is suspicious of God and wants him replaced, but it doesn't want to get caught. So, it sneaks about, looking for opportunities to oppose God's agenda and operating through subterfuge that avoids awakening the conscience. The flesh has a "Who, me?" attitude; it shrugs off responsibility. It becomes an expert in shiftiness and runs in the background like a deep computer program. Its purpose is to make sure that its operations will be bypassed. An unseen pattern of coping with life comes to dominate the human heart. What is the effect?

---

4. Leopold, *A Sand County Almanac*, cited in Wirzba, *Paradise of God,* 97.
5. Argyris, *Strategy*, 34.

CLOSED PEOPLE AND RELATIONAL DEBT

I mentioned God's agenda in the last paragraph, which, briefly stated, is to save and restore the world. When we believe in Christ, God establishes a relationship with us, the basis of which is a new covenant. In other words, we are saved by a new relationship, and its basis is love: "For God so loved the world that He gave His only begotten Son, that whoever believes in Him should not perish, but have everlasting life" (John 3:16). From the beginning of this new covenant, God has put a premium on how we relate to one another: "By this all men will know that you are My disciples, if you have love for one another" (John 13:35). But we can only love because "He first loved us" (1 John 4:19). That is, apart from our dwelling in God's originating love that creates in us a new disposition (to love God and love others), we do not know how to love; for what passes for love is too often curdled by self-interest. Earlier, we asked, "Where do we get the courage to face the question of our own contributions to relational deadness?" Where indeed, since, as has been trenchantly observed, "human kind / Cannot bear very much reality."[6] The courage to face the sins of our flesh and repent of their defensive routines comes from this newly absorbed love from God. If and when it becomes real to us that God loves us with a love that won't quit, the courage of self-examination rises. We begin, like David, to pray, "Search me, O God, and know my heart; / Try me and know my anxious thoughts; / And see if there be any hurtful way in me, / And lead me in the everlasting way" (Ps 139:23–24).

On the other hand, we are tempted to keep God at bay, even as Christians. We know that the Lord disrupts our preferred outcomes, saying things like, "Let the dead bury their dead" (Luke 9:60) and "Sell what you have and give to the poor.... Then come, follow me" (Matt 19:21). We give God lip service; but, deep down, we know that love calls us away from our selfish strategies. We'd rather ignore them, letting them drift down to the deep grottoes in our hearts. Bypassing such hidden realities not only flattens relationships (reducing their features and freedoms) but also closes us off from each other.

---

6. Eliot, *Four Quartets*, 14 (Quartet 1, lines 42–43).

CLEANUP

Earlier, we posed a question: When the flesh runs a hidden program in the heart, what is the effect? Since it's out of sight and undiscussable (bypassed), a person's true agenda is sealed from discovery, immunized to all light and freedom. The flesh, then, is a closed system; and its effect is to close us off from one another. By its very nature, the flesh develops patterns of safety and control, in effect forming a reliable, self-made fortress. From behind its walls, we make life cough up the outcomes we've decided are crucial for personal survival. Since this pattern of safety and control must be reliable, it has to be predictable. To be predictable, it must use maneuvers that have always "worked" to reduce pain. To increase predictability, the flesh resorts to the same playbook over and over, one that delivers that payoff of pain relief. The playbook becomes like a map we pull out time after time to remind ourselves that we're in control of life, that we have at our anxious fingertips those familiar routes of self-will, self-survival. Traveling those routes over and over wears a set of ruts that makes relational risk and newness more and more unlikely. Our survival systems become increasingly closed to new information, new possibilities, new risks. The self doubles back on itself in a self-concerned reading of its own feel-good gauges. The fallen heart's tendency is always to monitor its success in gaining the outcomes it wants. It ever twists back to its own self-interests and away from God and others. We end up enclosed in a small cosmos of self-concern, a tiny box of "me." After a while, we don't have to pull out the map anymore; we run on autopilot.

And then there's the unavoidable result that I, the self-protector, become less and less considerate of my being called to love you. I can't do both—can't protect myself and love you—because love is vulnerable or it's nothing. So, when push comes to shove, I will stay safe; I will maintain control. You will be unloved as I survive on my own terms. Already a debt begins to accrue, because the relationship is losing the nourishment that comes from love.

The flesh, then, is a closed system and makes for a closed person. One can't be an open person with a closed system of survival; the two simply don't go together—reliable, predictable survival comes from that playbook we keep under lock and key. We live inside this system that teems with strategies that close us to the

CLOSED PEOPLE AND RELATIONAL DEBT

hearts of others and to our own. Again, closed system equals closed person. Jurgen Moltmann's meditation on closed vs. open systems speaks profoundly here:

> Having called creation a system open for time and potentiality, we can understand sin and slavery as the self-closing of open systems against their own time and their own potentialities.... If [God] creates grace for the sinner, then he frees him from his self-isolation. We can therefore call salvation . . . the divine opening of "closed systems." . . . [T]he opening to God takes place through God's suffering over their isolation.... Thus man's openness to God is brought about by grace, and grace springs from the suffering of God in his faithfulness to isolated man [the self curved inward on itself].... Closed systems bar themselves against suffering and self-transformation [they are defensive]. They grow rigid and condemn themselves to death [the death of true relationships, for example]. The opening of closed systems and the breaking down of their isolation and immunization will have to come about through the acceptance of suffering. But the only living beings that are capable of doing this are the ones that display a high degree of vulnerability and capacity for change. They are not merely alive; they can make other things live as well.[7]

These are life words that open my heart. They speak of real hope that we can repent of those flesh-driven routines that isolate us from one another. By putting them aside, we repent of mere self-protection. We open up to the vulnerability of newness. The future becomes more than a robotic extension of the past. The map changes, reveals new territory, invites us to explore. Time doesn't have to weigh heavily any longer. Because we forge ahead to new territory, we're drawn into kairos time and away from mere chronos time. The latter is clock time, ticking away one moment like another. Chronos becomes heavier and heavier as the same old ticktocks pile up like tiny manhole covers on our souls. The former, kairos, is the time of opportunity, the time pregnant with potential, catalytic time.

---

7. Moltmann, *Future of Creation*, 122–23.

CLEANUP

When a basketball player takes advantage of a teammate's screen and is open for half a second—that's kairos. Our Father invites us to these moments, those in which our hearts are open to receive grace, awakening, hope, new life.

The flesh resents and fears this newness. It finds God's new map disorienting. In reaction, it demands the cocoon of clock time, the parade of the predictable. It's easier to control the known, harder to control the unknown. Yet control simply extends the past, making what should be a living future a limping repetition of yesterday. Tomorrow will simply recycle today. Addicted to reliability and predictability, the flesh is allergic to newness, dependent on uncreative routine.

This longing for the reliable and predictable gives us a new way to look at the idea of maps. Think of them as metaphors that jostle together the ideas of staleness and newness, sameness and openness. In other words, there are old maps and new maps. Is there a good scriptural analogue for maps? There is no actual map within the Bible itself, but there are several detailed descriptions of land and landmarks (especially regarding the promised land). Sometimes, the geographical descriptions have to do with boundaries, with the stretching out of expanded territory. At other times, the descriptions have to do with points of orientation (so-and-so was buried at the Cave of Such-and-Such, or the battle took place at Mount Something).

These descriptions order our perception of space (this happened here and not there; the promised land occupies this stretch of land and not that). Space is commanded and tamed through descriptive detail. Space tends to open when it is brought to order through the embrace of words (for example, words that might say, "To experience x, you will have to travel to point y or through territory z"). From the standpoint of Scripture, this ordering is always connected to God's playfulness, promise, and power.

In Scripture, then, these maplike descriptions depict the God who pushes against chaos, ordering space through word and promise, much as he did at the event of creation. Scripture doesn't use the word "map," but it does often use the word "path(s)." For example, Jer 6:16 says, "Stand by the ways and see and ask for the ancient

CLOSED PEOPLE AND RELATIONAL DEBT

paths, / Where the good way is, and walk in it." In the chaos of competing pathways, then, find the "good way" (that is, God's way) and pursue it. On the idea of chaos, Moltmann says, "Creation *ex nihilo* [creation out of nothing] is therefore creation *in nihilo* [creation within nothing] as well and is consequently creation that is threatened, and only protected to a limited degree against that threat."[8] After Satan's fall and the ensuing chaos, God's act of creating heaven and earth pushed chaos back but did not yet eliminate it (Gen 1:2). The fall of Adam and Eve compounded the problem, because chaos came into the good creation, riding on the back of sin. In rebellion against God's heart to preserve and protect his artistry, chaos surges against and seeks to deface God's beautiful works. Scripture invites us to "the ancient paths"—those that take us to new risks and new freedoms, those designed to restore creation and, therefore, the relational ecosystem. Those designed to restore love, in other words. The ancient path, we now see, is actually an invitation to a new map of life in Christ. New directions beckon us away from the chains of the flesh.

For example: I watch an adult and child at a coffee shop. The air between them is good. The boy calls the man "dad." The man points to something written on the coffee cup, and they laugh. They share a drink, a straw, a phone screen the dad is using. I think of emotional vitamins pouring from father to son, and I feel tears welling up. The father and son move me, because their love pushes chaos away. In the space of light that surrounds them, one soul is giving life to another, because death is looking for traction yet is sliding away, unable to touch them. The dad seems to have a path, a map of some sort. He knows he is giving gifts to the boy. Sin's mission to erase the traces of God's work finds little opportunity here. I see openness in this father-son bond. The father is open to his son, and so the son is open to the father. There is none of that closed-face relating that is so common and so cold. Their faces don't slouch away from each other in suspicion but lean toward each other in affirmation.

---

8. Moltmann, *Future of Creation*, 120.

CLEANUP

A map can signal that new territory lies ahead, that exploration and discovery are possible. This father and son are traveling by a good map. The dad seems to be reading it off his heart.

Two men across from me at the same coffee shop are poring over a literal map of the city that spreads around them. They seem to be discussing a route, trying to clarify the best way to cover a territory. They point, gesture. Maybe they're in sales. Maybe they're evangelists. Maybe they're looking for the next location for a new pizza franchise. Maybe they're city planners. Whatever they are, they're focusing the universe down to specifics: here is our route. They exclude every territory called "not this particular city." They narrow and limit in order to navigate and strategize. Similarly, the dad I described in the last paragraph loves on one boy at the moment and excludes all others. He routes himself through this life, this territory—this heart is the one he's pursuing, the heart of his son. Love focuses; it is not distracted. At the same time that he excludes other lives, the dad opens himself to this life across the table. Maps close some options in order to specify and open others. The chaos of opening to every life gives way to the order and risk of opening to *this* life, right here. To explore the new territory of another's life is to open that map, not a generic map or someone else's map or the map one prefers. When the two men opened the map to the city, they may have preferred another set of features; but those other features wouldn't have described that city.

On the other hand, maps don't always orient us, at least not immediately. Some maps have too much newness, combining exhilaration with confusion. Some maps are incomplete. Nora Pierce's novel, *The Insufficiency of Maps*, tells a story in which the gift of a map doesn't always solve the problem of reorienting. A character in the story uses different maps (not all of them literal) in an urgent attempt to find her home. Yet, the maps never quite steer her to the place of her yearning.[9] A new map, then, has the possibility of disorientation, inviting one into territory so new that one feels more vertigo than forward movement (like the areas marked "terra

---

9. Pierce, *Insufficiency of Maps*, 79–81.

CLOSED PEOPLE AND RELATIONAL DEBT

incognita" on ancient maps). Exploration and discovery can be a Tilt-a-Whirl of newness.

I think, for example, of the counselee with whom I've had a long discussion about moving from an old map to a new. His old map is crystal clear to him. It takes him through the following stops: scapegoat, fixer, prisoner, compartmentalizer, overfunctioner, shame-sponge, rager. This is familiar terrain. On the other hand, a new map is becoming clearer. It routes him toward freedom; it returns others' responsibilities to them, allowing them their own problems; it offers him healing, hope, humor, love. But as attractive as the new map is, it looks terrifying because the fallout from following its routes would be huge and, he thinks, not survivable.

I both understand his frozenness and am saddened by it. I grasp his fear, but I also sense his longing for the new map. He glances at it, is enticed; but the costs of getting there cause panic storms. His heart craves the old safety, and his flesh takes advantage of it, flattening his options and closing him off to newness. His flesh proposes, "Only I can help you, but you must obey the same old rules." But he's reasoning with deficient data, unable to recognize God's faithfulness as a factor enabling him to wrestle with the new map's risks. His old, well-worn version of shelter is like a familiar scent that seduces him with the lie that home is right up the road, just a few more steps. Here, you see—the old map shows the way! But it's a closed system, and it doesn't open out into a haven of freedom but instead curves back to the same old campground of familiar safety and control.

A few sentences back, I said the old map takes him through being a scapegoat, etc., but I should add the vital idea that this painful path also involves familiar experiences of seeming worth. Others count on his ability to fix their lives, to suppress his own feelings, to absorb, contain, and suffer through the unfinished business of the family system. He carries out these roles like a cleaner fish in an aquarium, vacuuming up detritus and keeping the tank (the family system) from poisoning itself. But he has to digest the poisons. Now we're back to the idea of a relational debt. He builds up a debt that the family system owes him, but the system won't "pay" by changing members' behavior. His body begins to pay the debt: he develops terrible headaches for no discernible medical reason; he's tired all the time; his hands fret over a numbness that comes and goes.

The brief experiences of worth conferred by the family system are satisfying enough to offset the pain of being blamed, feeling trapped, and so on. Why is there sort of a reverse "How much more?" that says, "Where pain abounds, the gift of worth abounds all the more"? But may we really speak of the gift of worth when such a dreadful price attaches to it? Are crumbs of worth worthwhile when they require that he suffer? The exchange of pain for acceptance is a form of salvation by works, and such a "salvation" simply makes one suffer the pain of hope deferred. And "hope deferred makes the heart sick" (Prov 13:12).

Suppose our sufferer prays for and finds the courage to turn, at least tentatively, to the new map. Why do I speak of the "fallout" that will come from his family? To explain simply: a weak limb loves its crutch. For example, if one's leg is stiff from an injury and one has come to fear the pain of physical therapy, a deteriorating feedback loop is set up. The increasing weakness of the leg increases the fear of the pain of therapy, thereby increasing the weakness, which increases the fear of therapy, and so on. By the same principle, the man afraid of moving to the new map ends up in a negative-feedback loop fed (1) by his fear of the new map and its vulnerability, and (2) consequently, by a deeper retreat into the protective habits of the old map.

Now we are back to the idea of closed systems. The negative-feedback loop consists, on one side, of that fear of the new map (and the relational fallout he dreads) and, on the other side, a further hardening into the familiar habits of the old map. On the fear-of-new-map side, the man looks for assurance that the old map works. Since he is so used to finding that evidence (say, when his father yells at him, he later gets a small benefit—Dad buys him breakfast, for example—and he "knows" everything is all right), it doesn't take much to accumulate feelings that the old map is reliable and predictable (the two words we used of the flesh earlier). Since he "finds" evidence that the old map is the "right" one, he ends up biased in favor of that evidence and filters the data of life through that bias.

CLOSED PEOPLE AND RELATIONAL DEBT

On the other side of the feedback loop, he ends up in a reactive hardening against the new map. In that hardening, he resists evidence that the new map works. For example, a friend affirms him, saying, "You don't have to keep suffering in your limbo; you have so much to offer in life." He immediately discounts the encouragement, saying to himself, "He's catching me on a good day. How can I have anything good to offer if so many around me think I'm a pain to live with?" As a result, he is biased against evidence that supports the new map.

The more one winds around this circuit, the more closed one becomes to the new map. The old map becomes increasingly resistant to the newness that God might want to bring. The evidence in favor of the new map is shed like rain from a poncho. Is there any hope of opening this system?

There's a "truism" one hears bandied about in the culture: people don't change. The truth is, they can and do. What the "truism" tries to express, however, is that it's easy to *resist* change, and many of us do. It would be better to say that there are reasons to hope for change, *and* change is difficult. One reason for hope is that as soon as we actually *see* that we're on a feedback loop, we have already taken a stance *outside* of it. Amazingly, the system is already opened a bit. It has to be if we're standing outside it, looking at our behavior! Standing outside the loop to get perspective requires a distance, like reverse Google mapping yourself to a vantage point *above* the loop. It is now in full view, or, as we say, it is now open to view. *Open* to view. You can only change what you can see. The unseen is unknown and unchangeable. But the seen, being open to view, is now not quite as closed as before. It now has the potential to disclose some of its secrets.

Since seeing is the first step to changing, many will not want to see and will maintain a chosen blindness: "What loop? What closed system?" Their resistance to change will dictate that they deny their own closed condition—i.e., they will be closed to their closed-ness. This grievous situation is reminiscent of Jer 17:9: "The heart is more deceitful than all else, / And is desperately sick, / Who can understand it?" But hope is in the next verse: "I, the Lord, search the heart, / I test the mind" (17:10). God has a commitment to bring

CLEANUP

truth to heart and mind. If we have a spouse, friend, child, or parent with a closed heart, Jeremiah gives us a place to stand and pray for that person. Of course, we hope that others are praying for us and any closing of our own hearts.

Again, the willingness even to see one's closed stance is a tremendous step. To clarify one's negative-feedback loop (biased toward old evidence, resistant to new evidence) makes freedom possible. A seeing person has opened up to the idea that he or she is closed! This person can now stand away from the old map and explore how it functions. This is an incalculable advance over one's being blind to the fact that a loop even exists.

This idea of function is the second idea for opening up the system, for unlacing the loop. If one knows how a system functions, one can divide the system into its parts. If one can divide a system into its components, one has a greater chance to understand its specific functions. Understanding can lead to awakening, which can lead to repentance. The components that operate the closed system are pain, fear, shame, flesh, control, and safety. A system can sometimes be changed by altering one part of it, thereby causing a disruption that destabilizes the whole shebang. We will look at these components one by one. The first we'll bring into view is pain.

CHAPTER 2

Pain (Fallen World 1)

Emotional pain can readily, insidiously become a tool of the flesh. Pain is an important "player" in the feedback loop we described a few paragraphs back. Moltmann, too, indicates something valid in this respect when he says, "The opening of closed systems and the breaking down of their isolation and immunization will have to come about through acceptance of suffering."[1] Why will "acceptance of suffering" have to be the portal for closed systems (closed hearts!) to open? The answer is that systems are closed against challenges that the person hiding in the system would find too scary. She anticipates intense pain if her system opens. For example, a woman who opens her map to new possibilities and actually leaves her abusive husband is almost guaranteed an avalanche of pain. On a rational level, she might claim with some accuracy that the pain outweighs the benefits, but her feelings may well say otherwise.

The dad who had a map of what to pour into his young son; the men with the city map spread between them, an orienting code that helped them to exclude so they could meaningfully include; the character in *The Insufficiency of Maps* whose pursuit of coordinates only left her lost in a new way—these vignettes teach us that maps can help yet may also end up being too much too soon. An overwhelming, new map puts one in touch with the possibilities

---

1. Moltmann, *Future of Creation*, 123.

of both tentativeness and vulnerability. New emotions, for example, can be equivalent to finding a new, disorienting acreage in one's heart. In my counseling practice, I see people suppress their tears with desperate self-restraint. "Why?" I ask them. "Because tears are weak," they say. "Tears are weak, how?" I ask. "Because they make you vulnerable." If pain is the reality we're avoiding, vulnerability is the treacherous door that lets pain inside.

How we hate vulnerability! A sitting duck—who wants to be that chump?! We'll desperately evade any map that taps into our fear of becoming the easy target. The wound of vulnerability is a wound with deep crevices, both theologically and personally. On the personal level, vulnerability is like placing one's heart right out on a table, unprotected. The heart on a table pictures an easily savaged nakedness. A heart on a table is a desperate question: "Will protection come?" The counselee who fights off tears doesn't trust that it will. The fear of pain dominates. The new map takes the heart into upsetting realms.

The closed system (or the old map) brings relief by shutting down the threat of vulnerability. The closed system eliminates risk from the get-go. It's as Carl Jung said: "Neurosis is always a substitute for legitimate suffering."[2] All closed systems, then, refuse to suffer the pain of newness. They are *The Wizard of Oz*'s Tin Man, only a stubborn Tin Man who wants no oil, no movement. The pain of unfreezing long-rusted joints is checked by the closed system. It's a prison where both newness and pain are locked away. There's a tragic trade-off: the Tin Man has no pain, but neither can he move.

Here's a little story about newness in chains. The story begins: When the counselor describes it, the listener can see the new territory. The counselor says, "It's a place of being real, taking risks, loving more vulnerably and more often. It's a place where God must meet and bring his love to you, or you will feel the threat is so high that you're dying. You'll feel exposed but more alive." The story acquires conflict: The counselor's words touch him, because he has wanted to be loved in that way. And he's never thought of loving in that way, of giving himself to meet the longing for love in

---

2. Jung, *Psychology and Religion*, pt. 1.1.3 ("The History and Psychology of a Natural Symbol"), para. 129.

PAIN (FALLEN WORLD 1)

others. He's seldom thought about their pain. That imbalance now strikes him as selfish. Why is he always pulling others to love him; why couldn't he just love others regardless of the return or lack of return? Enticed, he wonders what it would be like simply to reach for others and seek to do them good, moved by their hope for love? The story seeks to solve the conflict: but then he hears a voice in his mind, a voice calling his quest for newness weak and sissified. "People will think you're feminine!" sneers the voice. "How naïve! You'll just be walked over. You'll be used. Other guys will wonder what happened to you." The story winds down into sameness: the new territory doesn't look so attractive now. It mostly looks painful. To the counselor, he says, "You're going to get me a reputation for weakness and weirdness." The counselor looks puzzled, then says, "Jesus loved well and got himself crucified. Love can put you into positions that take a lot of strength." The man reflects and says, "Well, I'm not Jesus." The counselor responds, "It's not about being Jesus; it's about becoming the man he sees inside you." The response: "OK. Hmmm. This is just a busy time in my life. I have a lot going on." The counselor thinks about another man who said, "Let me go first and bury my father" (Matt 8:21), but only says, "Love will interfere with your life, that's certain."

From this story, we learn the power of pain to confine us, restricting our territory to sameness and warding off newness as out of bounds, scary, ill-starred. We think, "I'm doing okay in my comfort zone. It's a tad boxy, but it's stable, like a good Volvo." Pain continues to drive us. We don't let it teach us instead. Later, we will explore more about the tragedy of wasted pain.

First, let's look further at what makes vulnerability so difficult. We're shown what makes vulnerability unbearable right in the beginning, in the Bible's story of creation and fall. At the very roots of the human story, we find this: "And the man and his wife were both naked and were not ashamed" (Gen 2:25). No doubt, Adam and Eve were physically naked (as all the paintings assure us); but their nakedness also illustrates complete openness, one person freely exposed to the other, a self-revealing without fear. Sin has not yet arrived to prey on human hearts, so vulnerability brings no pain.

CLEANUP

Can you imagine? Neither man nor woman has experienced the wounds of betrayal, shame, mockery, failure, dishonesty, unfaithfulness. Neither has ever calculated an advantage over the other; neither has ever hidden a secret or had a self-serving motive; neither has felt the stab of broken trust, the stubborn darkness of self-will; neither has thought about needing to dominate the other or be subservient to the other. None of these common sufferings to which we are all so inured has yet intruded into God's protective creation, well-made to keep out the chaos. All is yet open and simple. There is no relational debt looming over the first couple. Their hearts are at ease, their nakedness a sign of complete harmony and tranquility. At the end of Gen 2, then, they stand naked, without trembling, with no sense of threat. They have come together in a relationship of pure delight, safety with no shame. Relationally, they are poised for endless self-giving: the dance of welcome and mutual completion. Textually, however, they are poised on a cliff's edge.

The original text of the Old Testament had no verses and chapters. What we think of as a break after chapter 2 (suggested by the heading "Chapter 3") simply did not exist. So, the flow of the lines would have been "were naked and were not ashamed. And the serpent was more crafty than any creature" (what is now Gen 2:25—3:1).[3] The pairing of human/naked and serpent/crafty throws a shadow over the clear air between Adam and Eve. Imagine the difference in the story's tone had the text said, "were naked and not ashamed. And the sheep was the meekest of all the creatures." The foreboding tone cast by "serpent" and "crafty" hints at the ruin to come. The text's uninterrupted move from harmonious couple to scheming snake spreads a pestilent fog over the radiance of creation. With the entrance of the serpent, it's as though a single, sad bell sounds, a bell of sorrow and warning. It is there in the undertone of a masterfully constructed text. Through this tense textual thickness, we sense the mobilizing of the predator.

The sudden transit from having nothing to hide to the menace of stealth and trickiness is underlined by the Hebrew words for "naked," describing the couple, and "crafty," which describes

---

3. My translation.

PAIN (FALLEN WORLD 1)

the serpent. The two words sound almost exactly alike in Hebrew (the basic words would be pronounced 'arôm [naked] and 'arûm [crafty]). We see that the author is punning here, not to be amusing but to emphasize tragedy. It's as though the writer means for us to see that the pairing 'arôm/naked with the nearby 'arûm/crafty suggests that malice (craftiness) hates innocence (nakedness) so intensely that it cannot allow innocence any breathing room. Evil is drawn to destroy good like a moth to a lamp, but more like Mothra than a moth. Evil's lust to ruin goodness makes for a never-ending emergency of the good (until Christ returns). Evil is not just a parasite on the good; it also has its own existence and is aggressive toward the good, because it hates the implied judgment that goodness brings. Says LaCocque:

> Abel before Cain embodies innocence and purity. It is precisely these virtues that offend Cain. Like Versilov in Dostoevsky's *The Adolescent*, Cain feels the urge to soil purity. Dostoevsky has perceptively unveiled the morbid desire in all the Cains of history to deface beauty. . . . Cain's killing of Abel is the paradigmatic onslaught against innocence.[4]

The serpent, being 'arûm/crafty, despises God's achievement, the goodness of God's creation. Repeatedly, "God saw that it was good" (Gen 1:10, 12, 18, etc.). The serpent's character shows evil despising the free air of innocence, hating the willingness of one person to do good to another. In a word, evil hates love, the relational form of goodness. In active hate, evil schemes toward the scouring of good from creation. Its agenda is the killing of love. The emergency of goodness and love brings a crisis of decision.

The decision amounts to a question: When goodness cries out for help, what will we do? Will we avoid the decision, clinging to our supposed right to define the good according to our own comfort zone? Or will we realize that life is not about how we are faring but about how the good is faring? To put it another way, life is about God's agenda, not mine. How far along am I in making that commitment? Do I trust God's agenda? Do I love that agenda? Or am I

4. LaCocque, *Onslaught*, 66.

CLEANUP

in love with my own list of outcomes, serving God with what I have left over after I service myself?

Here is the same dilemma clarified in Gen 3. The serpent presents Adam and Eve with a deal: step away from God and you can be like God. Yet, they already were like God, who had made them in God's own image! They were designed to image the Lord, to reflect his nature, and there had to be similarities between God and humans for that to happen (Ps 8:5). The serpent/Satan draws them in to a shell game where they tragically pursue something they already have. In asking the question in Gen 3:2, the devil appears ignorant: "Did God actually say, 'You shall not eat the fruit of any tree in the garden'?" (author's paraphrase). Of course, God only banned the fruit of one tree. LaCocque observes that "the serpent's obvious inaccuracy in his rendition of God's prohibition sounds like cunning or lack of subtlety. In fact, it is a well-known trick of the con-man to appear stupid to put others in a position of a sham superiority."[5] Part of the con game is that the devil constantly seeks to lure humans into a quest for something God has given generously already. Evil presents a crisis that isn't there by opening a fissure of doubt in the human heart. God, the one who dies for the good, is presented as the one who isn't good! And if God isn't good, he can't be trusted, correct? By this reasoning, we must trust ourselves.

The oldness of evil, its ancient hate of goodness—which is to say, of God—means that it can only *disguise* itself as new. But it can't be new, given that its predictable, uncreative, primitive, instinctive hatred of God constantly recycles the same strategies. As seen above, it reiterates the same old doubt about God. Newness comes only from God, who brings fresh, creative energy wherever he shows up. Brueggemann puts it this way: "If you celebrate what is, you won't receive what will be. If you are deeply committed to the old world that is now ending, you won't be present or available for the new world God will put into the void of creation."[6] With the word "void," Brueggemann refers to Gen 1:2: "And the earth was formless and void." A roiling chaos needed to be formed and

---

5. LaCocque, *Trial*, 145.
6. Brueggemann, *Interpretation and Obedience*, 317.

PAIN (FALLEN WORLD 1)

filled, and God, simply by speaking, exploded into it the newness of creation.

Self-sufficient, self-serving, self-protecting responses to the disintegration around us will always exude the oldness and tiredness of evil. It's the ancient, hateful spirit who says, "God has lied. God's prophets have lied. God's apostles have lied. The Son of God has lied. You can only believe yourself. You will not lie, not to yourself! Your heart is good, truthful, and healthy! Replace God the liar with your own heart's way. Believe in yourself. Trust your heart. God is an unnecessary complication. If God exists, he is a sneaky God with a hidden agenda. So, ditch all the smoke and mirrors of God talk and go your own way!" The flesh's lie is that, through our ingenious self-engineering, pain can be neutralized and managed. We make our personal utopia. So, who needs God? It is the oldest lie.

The promise that we can manage pain powerfully allures. We're enticed to believe in the recycling of evil's familiar solutions in new disguises. That's why God calls us continually to repent, in the crisis of decision, from our latest fall into evil. Be born again, again! I'm not talking about the possibility of losing our salvation here; I'm emphasizing the reality that new levels of discipleship need to be born in us every day. Sometimes, several times a day. Our urgent flight from pain means that we harbor hidden, unloving commitments that will require fresh expeditions of honesty for the rest of our lives: "Behold, Thou dost desire truth in the innermost being" (Ps 51:6).

Another reason that we must re-decide and re-repent: the lie that God can be replaced sets us up for a me-myself life. Käsemann notes that "such persons confuse their own wishes with the will of the Lord and dream of their own private gospel."[7] In this kind of life, we flee from pain on our terms. Determined to quiet pain at all costs, we assume that each of us must alone scrape together a life of quiet intensity—quiet in the sense that no one must know anything of each heart and its painful, anxious striving. In this lonely, frantic effort to squeeze drops of life from engineered outcomes,

---

7. Käsemann, *On Being a Disciple*, 71.

CLEANUP

we develop the illusion that we are free. It's not so bad, we think, that there's a basic loneliness to life, a deep alienation—it's not so bad because at least we're free to pursue our own self-designed outcomes. And this is a good freedom, right? Let's think about this alleged freedom.

In "Hebrew thinking . . . the world is in a constitutive relationship of descent and generation . . . . The thinker . . . is essentially the son of his father and mother (even when he ignores or hides it), and he prolongs willy-nilly, a text that had begun a long time before him."[8] By "text," Trigano means a story that is handed from one generation to the next and the next and the next. This idea of a handed-along story helps us think more deeply about freedom.

My own father, for example, certainly prolonged a text (a life story) that began with his maternal grandfather, who abandoned his wife and two young girls. That text, that story, was powerful in itself. It was made more powerful by the driving energy of my father's mother (the younger of the two girls who'd been abandoned). Her frosty anger and demand for control defended her against the pain of abandonment, but they conveyed a lovelessness to my father. That cold control enraged him and caused to push back both against his mother and his wife, my own mother. My father's tremendous grief in life revolved around the question "When can it be about me?" That is, "When will I be loved?" The text that predated his life and sought to predetermine it (and, in many ways, did predetermine it) threatened to make his story a mere epilogue of the "text that had begun a long time before him."[9] And, in fact, his life never escaped that story and did become, in the main, a mere epilogue. Can this really be freedom?

I'm using "epilogue" in its meaning as "a concluding section that rounds out the design of a literary work."[10] It conveys the idea of picking up an ongoing story and living it out it in a way that

---

8. Schmuel Trigano, quoted in LaCocque and Ricoeur, *Thinking Biblically*, 381.

9. Schmuel Trigano, quoted in LaCocque and Ricoeur, *Thinking Biblically*, 381.

10. *Merriam-Webster*, s.v. "epilogue," https://www.merriam-webster.com/dictionary/epilogue.

PAIN (FALLEN WORLD 1)

imprisons one within its dictates. One struggles yet fails to become free of the determining power of the preceding text that is too confining, negative, destructive. Does this sound like freedom?

One author refers to a "drive to repair" that inspires from time to time acts that "defeat determinism."[11] The same author says of Joseph from the book of Genesis, "Twenty-some years of envy, hatred, vengeful dreams; a score of years with tears, mourning, grudge-bearing, mistrust, bitterness are erased in one instant. Joseph has risen to consciousness and conscience."[12] This "one instant" refers to the moment when Joseph forgives his murderous brothers. Here, Joseph's repentance is illuminated and cast as a determination to be free, to break the bonds of determinism. Here, he frees himself (and his brothers) from simply continuing the text of a life predetermined by the story into which he is born. God being with him, Joseph rejects life as codicil—life as mere extension of the prior scripting that dogs all humans from the generational shadows. In the words "Joseph has risen to consciousness and conscience,"[13] we have before us the gift of realizing (if we will) that being biologically alive and functioning through a succession of passing years is not necessarily to be conscious and alive in a fully human way. It is not to be fully awake.

Continuing the story of unfreedom, my father's mother did not understand that she, in "writing" her life (making her moves to survive and pass through the years allotted her), was engaged in automatic writing, as it were. My grandmother survived by burying her heart's pain beneath layers of control, rigidity, and a forbidding propriety. Behind these walls, her deep sadness lay frozen. Instead of serving her and softening her, sadness drove her into an internal fortress. Her cool approach to relationships issued from the icebox of her tears. Frozen pain is a tyrant that enslaves a person to the story handed down from generations past. Avoided pain voids any contract with the future and hobbles one to the past. Can this be

---

11. Melanie Klein, quoted in LaCocque and Ricoeur, *Thinking Biblically*, 386.

12. LaCocque and Ricoeur, *Thinking Biblically*, 386.

13. LaCocque and Ricoeur, *Thinking Biblically*, 386.

freedom? Clearly, the lives I am describing are autonomous. But free they are not.

The metaphor of automatic writing conveys that life is not life if one merely "writes" what is dictated by one's history. Life as a prisoner of one's history drops into a lowland fog of blindness. It's no accident that LaCocque and Ricoeur use the metaphor of rising ("Joseph has risen to consciousness and conscience"[14]), for it communicates that Joseph's repentance requires an awakening, a departure from the bog of powerful, determining drives for revenge, hate, and control. He repents through a resurrection of sorts. He becomes "erect, face turned upward to God in a listening-speaking relationship."[15]

Leanne Payne's picture of standing tall, "face turned upward to God," returns us to the struggle to move repentantly from self-sufficient to God-dependent responses to life. I'm using the term "self-sufficient" to convey the attitude that "I can make life work on my own terms. I can pound on life to get the outcomes I want, outcomes that make me feel safe, gratified, and in control." By contrast, I'm using the term "God-dependent" to describe a heart that enters into disappointed longings with honesty to the point that if God is not God, one will die from starved longings. Again, David calls this "truth in the innermost being," a condition God longs for us to embrace (Ps 51:6). This begins to sound like freedom. How so? Because honesty about disappointed longings liberates us from self-deception. We no longer have to run from the truth that we are born into a frustrating world, a world that repeatedly hurls us helplessly into hurt.

For years, I handled the onslaught of unfulfilled longings by steering (as if from a storm) into the short-lived shelter of false relief. In my early days as a believer, I kept right on seeking such "harbors" as pornography, masturbation, food, the addicting salve of approval at all costs, a contrived passivity that gave me a free pass from life's challenges. Instead of crying out to God, "Mentor me; lead me to be a man! I doubt myself, but, Lord, come to my

---

14. LaCocque and Ricoeur, *Thinking Biblically*, 386.
15. Payne, *Healing Presence*, 52.

PAIN (FALLEN WORLD 1)

doubt and carry me through," I pursued these false loves that required little and soothed my fear with fancies of fabricated comfort. But there was no true comfort, only addictive cycling into greater and greater shame, a deeper retreat from life, and a burdening of my marriage with my polite but stubborn demands. That was my closed system at the time.

In later years, the temptations have been different—e.g., to flag and fade into fatigue and discouragement, to become cynical (because evil seems to get the upper hand in the world all too often), and to slide toward self-pity. All these disappointed longings! Why does evil seem to win so often? I'm tempted to kill my longings to soothe my hurting heart. This is my current temptation to live in a closed system. Again, if God is not God, I will die from starved longings. Or from longings I've murdered. The heart is really alive only when it longs without demanding satisfaction, yet lives in hope that God is the water of life. Only a heart filled with unfulfillment yet crying out to Jesus is a heart turned toward the living God. Only a heart brimming with unstifled longings is truly awake. Managing life with the deftness of self-will suffocates our real selves. "Truth in the innermost being" requires that we say, "My self-driven solutions to life's pain only lead to foolishness in me and heartache for others. I damage others in my drive for demanded outcomes. I confess my sin, O Lord. I throw myself on your mercy, and I relinquish outcomes to you." This is the God-dependent life. This begins to sound like freedom.

The God-dependent life is a life of nakedness. Exposed to the full impact of unmet yearnings, the naked heart glances toward tempting alternatives easily found in the self's storehouse of broken cisterns (Jer 2:13). But the God-dependent heart only glances at that foolish supply and then looks searchingly to God. Earlier, we saw that Adam and Eve "were both naked and were not ashamed" (Gen 2:25). Encountering one another face to face, body to body, heart to heart, mind to mind, soul to soul, the two had nothing to fear. Vulnerability, in those ancient days, was precisely invulnerable in that the shock of shame and sin had never entered creation. Pristine and beautiful, creation and Creator experienced free communion,

CLEANUP

and within that communion, the union of male and female was sheltered from all hazards (except the misuse of their freedom).

They did abuse their freedom: now, relationships are full of hazards such as manipulation, deceit, betrayal, rejection, power games, abuse, shaming, slights, and stings. This is not to say that relationships can't also move toward mutual respect and love; in fact, some do. But relationships are now risky, and vulnerability is more like a load to carry than a freely chosen, mutual dance. Why take such risks? The answer takes us into territory that may not be familiar to North American minds—i.e., minds used to thinking in instrumental terms. By "instrumental terms," I mean the stance that everything that is not me is something for which I might find a use. We tend to think of things—and people—outside our own skins as instruments, and we tend to calculate how they might (or might not) be profitable to use. The category in which we're *not* used to thinking is that of true otherness.

"Thus the fisherman gets his catch. But the find is for the diver," says Martin Buber.[16] In this profound aphorism, the fisherman treats the lake as a mere fish tank, a resource, an extension of his toolbox-for-life-management. He brings the tools he already has (rod, reel) to extract fish from the lake. The lake is a tool for containing fish: everything is merely useful. The diver, on the other hand, is fully involved in the lake. Notice the "in" in "involved." The lake is a world, and a world can only be known through entering *into* it, exploring it, being in it. The diver is immersed. Given enough time, she can discover the lake's secrets. She sees how the sunlight filters through in glowing filaments, turning amber as they slant down and down. She is curious about an algae bloom near the shore and finds the culprit: an unregulated sewer pipe. She sees how freshwater clams dot the shallows, and she wonders how they compare to their saltwater kin. Are they edible? How would they taste in chowder? Curiosity tickles her to thought. Four feet down and glancing upward, she sees the plunge of an osprey as it lances a fish. She thinks, "It's like a fallen chandelier scattering a

---

16. Buber, *I and Thou*, 55, n. 4.

PAIN (FALLEN WORLD 1)

galaxy of fugitive shards." There is nothing useful here, just "radical amazement"[17] with its invitation to beauty.

"We either contemplate, or we exploit," says Alan Jones.[18] The fisherman exploits the lake for its finny contents. The diver contemplates the lake as a living world of changefulness that responds only to attentiveness. Since attentiveness takes time, and since we live in a world that rewards speed, there is little real attentiveness in this world. It seems our whole culture suffers from a deficit of attention. One could argue that people diagnosed with attention deficit disorder are those with more severe cases of inattentiveness in a society where being easily sidetracked or distracted is pervasive. We simply scan over most people and most things. We barcode everything for its estimated usefulness. Once we've sized it up, we either use it or move on. Sizing up and moving on are not the ways to discover otherness.

Otherness is the scandal of something or someone being not-me. If I stop scanning and start paying attention, I see millions of bundles of buried promises and challenges. Each person hides a cluster of potential and stuck-ness. If each person is a burial ground of the two, then the spade is attentiveness. Paying careful, "time-wasting" attention brings otherness near. I begin to realize that each person is a world, and I'm using "world" to mean a bounded, distinct set of peculiarities usually hidden behind a culturally acceptable mask. As with the lake, no world can be known without full immersion.

As we engage this idea of otherness, three great themes emerge. First, otherness is present whenever something not-me is present. This idea implies that if I pay attention outside myself, I will see otherness more and more clearly. The second theme is that the more I get used to otherness, the more I can stop thinking of the things and people around me as commodities. Third, otherness can only be appreciated through immersion. How do we immerse ourselves? Through quiet, steady, focused attention. This is a state of mind easier described than accomplished. For one thing, attentiveness

---

17. Heschel, *Man Is Not Alone*, 30.
18. Jones, *Soulmaking*, 29.

CLEANUP

means being still, and in Western culture one who is standing still and paying attention (say, to a skyscape) is likely to be asked, "May I help you?" with an implied anxiety. Yet, I see no alternative. If we are going to immerse ourselves in otherness, attention is the only way. Inattentiveness, scanning, barcoding the people and objects around us—these are simply our unmindful habits of classifying everything outside us as resources, threats, or insignificances. Only when we stop thinking in such categories can we see that each thing or person is a world—like the lake in our example of the diver—for exploration and understanding.

Whatever has engaged our attentive diver has charged her heart and alerted her to a vast swath of things-not-myself crying out to be known. What does this mean? A soul that has withdrawn its attentiveness is at least on hold, if not dying. I hasten to add that all souls are attentive, but we too often attend urgently to the whole project of managing pain. The result is that our attention is directed inward toward the invisible neural blanket of our defenses. Constantly reading the state of those "nerves," many of us exhaust our attention, since we are finite and cannot pay heed to everything. The question is not whether we will pay attention, but where. When my attention homes in on reading the gauges of my pain-management efforts, I'm simply too preoccupied to "go jump in the lake"—that is, to stay on the track of otherness.

CHAPTER 3

Fear and Shame (Fallen World 2)

Pain dogs our steps in a fallen world. We're not home, and it hurts. Living with the internal ache of inconsolable homesickness reveals that life is bigger than we are. The task of consoling the inconsolable is so daunting that it touches a nerve, a deep concern that we are in over our heads. The sheer size of life and the search for consolation induce both fear and shame. The fear is that we won't find comfort. The shame is that others may find out about our fear.

FEAR

Clearly, pain is a major player in shutting us off from others. Specifically, fear of pain has the effect of nailing one foot to the floor, relationally speaking. We keep pivoting around the point of making sure of our own safety. We ask ourselves, "How quickly can I return to base?" as if we're in one of those childhood games, like "kick the can." "Base" is our closed system of safety and control. Instead, we could be asking, "What does it mean to do the other person good?" We may end up far from "base" but closer than ever to a now much needed God—much needed because we have entered the vertigo

CLEANUP

of allowing otherness to be real. When we focus on the other, we realize he or she, too, calls out for comfort.

How can we transform pain from a limiting to a liberating reality? Instead of our seeing pain as a driver, we can begin to see it as a teacher. When the flesh is able to hijack our pain, making us feel alone in it, we feel the emergency of being alone, of being "voted off the island." How does the flesh "hijack" our pain? Gal 5:13 tells us, "For you were called to freedom; brethren, only do not turn your freedom into an opportunity for the flesh." In the words "opportunity for the flesh," we learn that the flesh is an opportunist, that it always seeks an opening to steer us away from God and toward our self-interests. In Greek, the word "opportunity" means "base of operations."[1] It might describe an army using a base of operations from which to attack. Similarly, the flesh uses pain as a base of operations from which to attack us with ideas of aloneness, abandonment, terror. When pain hits, the flesh pipes up and says, "God doesn't care about you. You are on your own like everybody else. Don't be a loser and get caught with your defenses down; get busy and master the pain." This is what I mean by "the flesh hijacks our pain." Once the pain is hijacked, it drives us, and the more it drives us, the more we forfeit its ability to teach us.

Fear of pain tempts us to keep our coping systems tightly closed. Jesus, addressing us as fearful beings, says repeatedly, "Do not be afraid" (e.g., Matt 14:27; 17:7; John 14:27). Why does he need to reassure us so? What are our fears made of? We are afraid because we are capable of being injured. We are pierceable by many things, concrete and abstract. Sticks and stones can break our bones, *and* words can harm us dreadfully. We have no carapace against either physical impact or emotional stings. We are reduced to coping, and coping is tiring. Coping works like the shields in *Star Trek*. The Enterprise protects herself from, say, photon torpedoes, but each hit drains the shield's power. Eventually, the shield, depleted, fails. This is the result of coping; it works, but every act of coping drains one's battery, so to speak. The drawdown of potency makes us afraid. When will we crack?

---

1. Abbott-Smith, *Greek Lexicon*, 72.

FEAR AND SHAME (FALLEN WORLD 2)
SHAME

Shame is another important player in keeping us within our closed systems. Shame shuts us into ourselves in its concern about being exposed, being caught out in the open by critical eyes that turn into mocking, rejecting voices. Eyes of contempt easily become voices of disdain. Shame is also despair over wishes that don't come true. We feel ashamed when we wish for a world full of love and support and then realize we live in a world rife with betrayal. We feel ashamed that we trusted, or wished, in the first place that life could be a paradise. What kind of fool makes such a wish?

Shame, then, could be summed up in two ways: "I feel naked and embarrassed" and "I feel foolish that I hoped." Kaufman uses the picture of a trust bridge to illustrate shame. He says that we come to expect a caring response from another, a two-way street, a "mutuality" of respect, and when our expectation isn't met, the trust bridge is broken. That breakdown is a "potent generator of shame," according to Kaufman. "What a fool I was to trust him!"[2] The heart, he says, is exposed to view and feels stupid for opening itself up to trust.[3] The feelings of nakedness and foolishness tempt us to retreat or to attack. Either reaction reinforces the pain of life and pulls us toward justifying our coping system (driven by the flesh) and closing it even tighter. The flesh hijacks our shame and tells us a plausible lie: "This terrible feeling of shame should convince you that life is too risky; relationships are too scary. All you can hope to do is avoid shame by protecting yourself. Self-protection has to be your North Star, your governing directive for life." The more this mantra takes hold, the more we retreat into our old map for coping with life. The same old procedures dominate us, and change is farther away.

A recent counselee found her life shattered when her husband had an affair with a younger woman. The unspoken message she got was "You are old, substandard, unwanted, no fun, a drag. You have nothing to offer." Shattering her trust, breaking the bridge of love on which she'd felt secure, the affair destroyed the confidence

---

2. Kaufman, *Shame*, 14.
3. Kaufman, *Shame*, 14.

CLEANUP

she'd felt for twenty-five years while raising three children. Shame whispered, "How could you have been such a fool? You should have seen it coming. You missed the signals. What is wrong with you? You're such a loser!" Dripping caustic lye into her soul, these agonizing thoughts dragged her farther and farther down. She began to wake in the middle of the night with an oppressive weight in her chest as she thought, "Who is going to want me? I'm totally unmarketable." Life had dropped her brain into a hornets' nest. Her mind felt like an anvil. How powerful is shame? Powerful enough to drive a depression.

How do we convert shame from pure toxicity to an emotion with some hope of redemption? We need to see the longing at the heart of our experiences of shame. To get at this idea of longing, we can turn to Isa 55:1–3:

> Ho! Every one who thirsts, come to the waters;
> And you who have no money come, buy and eat.
> Come, buy wine and milk
> Without money and without cost.
> Why do you spend money for what is not bread,
> And your wages for what does not satisfy?
> Listen carefully to Me, and eat what is good,
> And delight yourself in abundance.
> Incline your ear and come to Me.
> Listen that you may live.

God appeals to us from his heart: "I am the water you are searching for." Through God's passionate voice, he reveals how painfully thirsty we are, how filled we are with longings. We're created to have nourishment that doesn't require anxious striving. We're not designed for mercilessly whipping ourselves to new heights of performance that we can trade in for "water." God quenches our thirst with "wine and milk *without cost*" (Isa 55:1, my emphasis). In other words, God brings us gifts and is the gift.

Experiencing shame is the opposite of receiving a gift. Just think of someone's bringing you a gift: you are out in the open in a delightful way. You are exposed—not to the acid of shame but to the refreshment of being enjoyed. There is only blessing, nothing tricky or costly to it. There is only generosity. In shame, on the other

FEAR AND SHAME (FALLEN WORLD 2)

hand, you feel as if you've received a nasty and unexpected bill. Now, you're going to pay. Caught out in the open at a disadvantage, you're exposed, and others will know how deficient you are. You'll writhe with embarrassment, abandonment, loneliness, nakedness, mockery.

We'll pay almost any price to avoid such drastic exposure. But avoiding shame is costly. It's as if we're stuck between two walls closing in: one wall is the high cost of feeling shame; the other is the high cost of pretense, hiding, fake "okay-ness," distancing, fancy footwork. Through these stressful efforts, we monitor our perimeter with anxious care. Whether we're avoiding shame (through pretense) or actually feeling shame, the price is high. This high-cost world is not the world for which we're designed. We are created not for anxious strivings and uncertain earnings but for giving and receiving gifts. Adam, for example, wakes up from being created into a dazzling world of color, texture, allure—a symphonic tonic of astonishing sounds, a sensory bath: gift after gift after gift. Adam has gained the whole world, a whole soul, and is about to receive the gift of Eve, an equally whole-souled, wide-eyed, trying-to-take-it-all-in heart of wonder.

How do we heal from shame? Gift and wonder help a lot. Longing, which I mentioned earlier, tying it to thirst (Isa 55:1–3), connects deeply with gift and wonder. How so? Back to Adam and Eve. The fact that they awaken, newly created, into a flood of impressions most sensual, delightful, artful, beautiful—this superabundance of incoming enchantments implies that the core of being human is the longing to live saturated with beauty, drunk on beauty, rolling around in glory. The relational form of beauty/glory is love. The poetic form of beauty is creativity, especially that which increases the sum of good in the world. So, we might say that humans have at their core dual longings, those for love and those for creative acts of goodness. Longing for love, longing to tilt the scales creatively toward goodness—these are our God-designed thirsts.

Another way to think about the word "longing" is to chop it down to the root, "long." Unsatisfied desire inherently involves waiting, and the waiting feels long, no matter how much time elapses before fulfillment. It puts one under strain, the strain of enduring.

CLEANUP

Take thirst, for instance. True thirst feels agonizingly long, even if it only lasts five minutes. Longing is about the question "Will it ever get here?" It seems to take too long. And so we wait. A long time. To exist as human is to long, to yearn.

In a fallen world, our waiting (for love and for goodness to win) runs up against Satan, who has come "to steal, and kill, and destroy" (John 10:10). The evil one is the thief who steals the fulfillment of our longings, of our good desires for love and for the victory of goodness (which is to say, for God). The devil shames our longings, intoning, "It's weak to yearn for love. Supermen/-women move above the fray. You are pitiful in your neediness." He takes advantage of our embarrassment by offering us less risky fare that gives us short-term relief (to wait for love is vulnerable, but to look at porn is not). And the fall into short-term relief brings us long-term loss. We trade our birthright for a bowl of soup.

Psalm 40 begins thus: "I waited patiently for the Lord" (40:1). The result of this waiting is that "He put a new song in my mouth, a song of praise to our God" (40:3). Patience quiets the heart, enabling one to send the taproot of trust into the soil of God's potency: "He brought me up out of the pit of destruction, out of the miry clay" (40:2). God's saving love is the basis for the new song of praise. Patience sets up a theater for God's incoming power, and his power signals to our longings that healing is in the offing. Instead of shaming our longings, God encourages them, telling us often, "Wait for the Lord" (Ps 62:1), and waiting *is* longing. We can expect from God not shame but a story in which his pursuit of us wins out over the one who comes "to steal, and kill, and destroy" (John 10:10). Patience, we can now say, makes for a kind of longing that leads not to sin but to hope. Since "hope does not put us to shame" (Rom 5:4 ESV), patience is a strong antidote to shame.

What we've seen is that shame tempts us to close our coping system against hope and to stay sealed in our ways of surviving. But the reality of longing and the leavening power of patience in our longings combine to open us to God and his good story. In that story, God calls us to love and be loved. By the work of the Holy Spirit, we are caught up into Christ's opening the door to the throne of grace. From there, the Father welcomes us to the banquet of

FEAR AND SHAME (FALLEN WORLD 2)

well-being, killing "the fattened calf" (see Luke 15:23) and inviting us to rejoice at the festive party. We find a release from penury and pain, heartache and hopelessness. We dance. Self-consciousness? Shame? Increasingly, they become things of the past.

Now open to God, we can open to others in truth and in love (Eph 4:15). John of the Cross describes a God who "takes your hand and guides you in the darkness, as though you were blind, to an end and by a way which you know not."[4] Pushing beyond our comfort zone invites us to take unfamiliar paths. Progressively free from shame, we hear God's call. The Lord allures us with his promise of "grace to help in time of need" (Heb 4:16) as we expand our options, choosing the risks of doing good to others in vulnerable connecting. Recently, a counselee spoke to me about how he works hard to compensate for a poor self-image. He puts an ultra-competent persona out front and makes sure he can justify his behavior; he is quick to defend himself against criticism, no matter how well-intentioned or constructive. Inside, though, he feels like a toy that came off the assembly line already broken and defective. No one must suspect such brokenness, so he works furiously at competence and defense. A respected doctor, he pulls off the agenda of competence, but he deeply hurts relationships in the process. He can't be real, and realness nourishes true connecting. It was a great day when his wife gave him some honest feedback, and he did not defend himself. Instead of construing her honesty as a threat ("pushing me toward the abyss," he used to say), he opened himself to the mirror she was showing him, a mirror that showed his false, distant style of relating. It was a revelation; he saw himself as a selfish survivor who had put a great burden on his marriage. He realized he was shifting onto his wife the load caused by his defenses, and she was staggering under it. He went, in a moment of grace, from being a closed to an open system; and he opened "the eyes of his heart" (alluding to Eph 1:18). His wife felt heard, validated, loved. His openness made him available to her; after many lonely years, she could come close to him. His repentant response illustrates the awakening described in one of the great passages in literature:

---

4. John of the Cross, *Dark Night*, 154.

CLEANUP

Mr. Head stood very still and felt the action of mercy touch him again, but this time he knew that there were no words in the world that could name it. He understood that it grew out of agony, which is not denied to any man and which is given in strange ways to children. He understood it was all a man could carry into death to give his Maker, and he suddenly burned with shame that he had so little of it to take with him. He stood appalled, judging himself with the thoroughness of God, while the action of mercy covered his pride like a flame and consumed it. He had never thought himself a great sinner before but he saw now that his true depravity had been hidden from him lest it cause him despair. He realized that he was forgiven for sins from the beginning of time, when he had conceived in his own heart the sin of Adam, until the present, when he had denied poor Nelson. He saw that no Sin was too monstrous for him to claim as his own, and since God loved in proportion as He forgave, he felt ready at that instant to enter Paradise.[5]

The freedom of repentance wreathes Mr. Head in gratuitous beauty. Our hearts beat faster here, because we are made in God's image. We are created to be like Christ, who, as the true human being (yet very God), shows us the way ("I am the way" [John 14:6]) to reflect God in how we live and relate.

Thinking further about this man who repented, we hit upon a surprise: his openness emerges in the midst of pain. A great learning unfolds here: instead of running from pain, we begin to lean into pain. As we turn to pain to search for its messages, we grasp that within pain are hints and hopes of aliveness. Earlier, we spoke of converting pain from a driver to a teacher. Instead of running from it, we dare to believe that pain is a loud alarm insisting that we are not yet home. Is this a valuable lesson? Unbelievably so. If we are not at home on this planet, we can begin to let go of the pressure to make life work here. The fact is that life here does *not* work well. How could it? We live in a fallen world where everything shares in a basic brokenness. Life has "its own obstinate shape that we cannot

---

5. O'Connor, "The Artificial Nigger," in *Complete Stories*, 269–70.

FEAR AND SHAME (FALLEN WORLD 2)

iron out to our liking."[6] We play "Whack-a-mole" with life, trying to get everything to stay in place; but problems pop up faster than we can solve them. If we don't listen to the pain this causes, we end up working so hard to whack the "moles" that we hyper-focus on control and have little left to give others. Relationships suffer.

As we let it sink in that we are not at home here (we are "strangers and exiles on the earth," according to Heb 11:13), we gain the ability to translate pain more accurately. Instead of interpreting pain as something to avoid, we read pain as telling the story that humans are achingly homesick. Recently, I was reflecting on some losses in my life. Outdoors doing some yardwork, I felt a gloomy ink squirting into my brain. The whole day became darker. The hurt in my soul amplified until I thought I'd crumple to the earth with the sheer heaviness of loss and grief. I came to a fork in my emotional road. I would either descend into hopelessness and the outer edges of depression, or I would find a way to retranslate the pain. I cried out inside, "It's only pain!" By that I meant something like "As horrible as I feel, I am not in the presence of something lethal. Many have felt great pain in life and profited from it. I don't have to let it intimidate me. I can pray to my Father 'who sees in secret' (Matt 6:4), trusting him to remind me that there's something beyond the pain." Slowly and with difficulty, the pain transformed itself into a language. It was as though suffering had arrived to clarify a struggle in my heart. Here is what I sensed: "You are demanding an outcome that you simply can't bring about. It has become too important to you. You are wounding yourself by gluing yourself to a false hope, an idolatry. You're afraid to let it go, because you think the loss will kill you. Without me, it *would* kill you. But I am here. You don't trust me enough, but I will help you in your unbelief." Slowly, a trickle of aliveness began. The hope came that God brings a perspective that makes the universe of pain smaller. As it shrinks a bit, we realize there is someone beyond the pain. Not only is God outside the pain, he is outside the universe yet turned toward it in redemptive love.

Perspective comes as we regard the pain as carrying a message. I had to translate the pain into that message. Pain cannot

6. Kidner, *Time to Mourn*, 15.

be transformed unless it is translated from wordless hurt into the world of discourse. That means personal pain can become part of a great conversation. It can change from a mute internal scream and become part of the world of ideas. Fine if it starts as incoherent speechlessness, an animal screech in the heart; but it doesn't have to stay there. Pain is pregnant with words, and one's keeping it mute is a stillbirth that means recycling the pain over and over until it kills hope. Pain, then, becomes perspective as it becomes speech. Pain becomes a way of seeing when it becomes a way of speaking. Honest speech gives us freedom to talk our way into seeing. When the heart speaks its pain, the person sees—new light comes. In the moment I described above, when I brought my pain to speech, I saw my idolatry. Now, I am carried into a whole world of discourse about idols and what to do about them. Again, pain-become-speech means new light, the defeat of shadows.

Take the Psalms as an example: Walter Brueggemann sees them as paradigmatic examples of bringing pain to speech. He says:

> The issue that Israel and Israel's God ... must always face concerns pain—whether pain is simply a shameful aberration that can be handled by correction or whether it is the stuff of humanness, the vehicle for a break with triumphalism, both sociological and theological. What we make of pain is perhaps the most telling factor for the question of life and the nature of our faith.[7]

What Brueggemann calls the "break" of embraced pain I would express as translating pain as one would a language. In Brueggemann's thought, such translating breaches the paralysis of dumb suffering by giving a new account that requires expanding one's formerly rigid view. When I expressed my internal pain in words, I moved from "This pain will kill me" to "This pain is telling me about a struggle I'm having, a struggle to let go of an idolatrous demand." To cite Brueggemann again: "Everything must be brought to speech, and everything brought to speech must be *addressed to God*, who is the final reference for all of life."[8] As I bring pain

---

7. Brueggemann, *Old Testament Theology*, 19.
8. Brueggemann, *Message of the Psalms*, 52; his emphasis.

FEAR AND SHAME (FALLEN WORLD 2)

to speech and address that speech to God (that is, as I pray), my thinking begins to stretch around the problem of letting go. I enter that world of discourse where I ask questions like "What is hard about letting go of this idol? How have others let go of hoped-for outcomes? What is on the other side of letting go? Are there any examples of letting go that I can find among my friends, in literature, in Scripture? Can I form a community of letting-go people? What would God have to say about this letting go?" Again, I enter into an entire world of ideas and relationships that provide help as I face this ordeal of letting go.

CHAPTER 4

Flesh (Anesthesia)

In that last sentence, I use the word "ordeal" purposefully. Letting go of idols is a severe trial that requires grace from God, grace that overcomes the flesh. We have already seen that the flesh wants to close our individual coping styles around the governing value of self-protection. The heart of the flesh is pride that expresses itself thus: "God can be replaced. God is a painful presence who just wants to command us and spoil our lives with dreadful assignments, sacrifices, and toil. He is not to be trusted. Life is about trusting myself and developing the resources to get the outcomes I desire." Ezekiel 28:2 captures this attitude well: "Your heart is lifted up, / And you have said, 'I am a god, / I sit in the seat of the gods.'" And Isa 14:13–14 goes farther yet: "I will ascend to heaven; / I will raise my throne above the stars of God, / And I will sit on the mount of assembly / In the recesses of the north. / I will ascend above the heights of the clouds; / I will make myself like the Most High." The human heart, left to itself, devolves in this proud direction, constantly grasping life to itself with the cry, "Life should go my way. How can I pound on it harder to make it come out to my liking?"

Few of us have the guts to reveal our hidden pride so clearly, so we deceive ourselves with a veneer of civility, including Christian civility. The "civil" way to describe and justify the flesh could sound something like this: "I know I'm not perfect, but I try hard to be a good person and a good Christian. I try to follow God, and I pray.

FLESH (ANESTHESIA)

He says he will give me the desires of my heart. There's a way to live life where he gives the guidance and answers I need. I just trust him and keep doing the next thing." It's not so much that these words are terrible in themselves, but so often they're enveloped in a tone of mastery. The impression is that one has a firm grip on life and is managing pretty well. The real "god" in this scenario is the managed life. But life in a fallen world is not manageable. Because the world is broken by sin, too many unpredictable contingencies come over the horizon for us to manage them. You can't play "Whack-a-mole" on an acre-sized board, no matter how quick you are. God calls us to renounce the misshapen idea that the key to life is personal resourcefulness, including the resourcefulness to bring him in on the whole life-management project. This is simply self-centered. God ends up being merely one more tool in the box for making life come out right. He might be the brightest tool and the one used with the most respect, but a tool nonetheless. God has been reduced to usefulness.

God never submits to being our tool. He knows how prone we are to instrumental thinking—that is, that everything that is not me is an instrument that could help me climb to happiness or guard it for me once I get there. God's desire for us is that we look beyond the limit of seeing life as a craft, a set of skills to be mastered. If life is only a craft, then it's just one more performance to be achieved, one more arena in which to be judged or in which to develop arrogance. If life is only a craft, I say. Life *is* partially a craft and should be approached as such, while at the same time, life is an unmanageable plunge into the vortex of a fallen world in which the point is not to craft one's way through the vortex but to accept the pain of it as revealing God's desire to open the "eyes of [our] hearts" (Eph 1:18). When we open those eyes, the epiphany is that life is not primarily a task but a gift. How a gift? The idea is that we're born into a condition where we lack life. Oh, we're very much lively as infants (normally) and move about, learn, grow, develop, excrete, eat, etc., etc. But this is biological life more than it is true animation (the word *anima* in Latin means "soul"). Real life emerges as the soul reconnects with the source of life, the God of life (so often called "the living God" in the Bible; see Pss 42:2, 84:2).

CLEANUP

We are born, then, deficient of life. We are born, in a real sense, dead. Paul conveys this idea when he says to the Ephesian Christians, "And you were dead in your trespasses and sins" (Eph 2:1). Since sin separates us from God, and God is the true source of life, we are cut off from true life; we live a depleted life. Nor should this be too surprising, since God told Adam, "In the day that you eat from it [the forbidden fruit] you shall surely die" (Gen 2:17).

Through our coming to believe in Jesus, God "made us alive together with Christ" (Eph 2:5). The adversary of life, Satan, despises our coming alive, especially the fact that once we believe, he cannot change our newly alive status. On the other hand, it is his nature to oppose our living out the new life God has given us. He reloads and comes after us. Having lost us to new life, he labors to keep that new life from full expression. I imagine that Satan has nightmares about a bunch of free Christians actually living as "little Christs" (which is one way to think about what "Christian" means). He opposes us through temptation, accusation, despair, deception, and attack. This book is not the place to develop a theology of spiritual warfare. For starters, you might want to read Richard Lovelace's *Dynamics of Spiritual Life*, especially his section "Authority in Spiritual Conflict."[1] For our purposes, the main point here is that we are born in a state of real deadness. It's not just *as if* we are dead—we really are dead in respect of true life from God, who is life. It is this deep sense of our own actually present death that the flesh uses to stir us to run from pain, because pain always carries a tinge of death. Nevertheless, God pursues us. When we turn to the Lord in faith, we actually come alive: "If any person is in Christ, that one is a new creation" (2 Cor 5:17, author's paraphrase). Yet, the newly created person lives in a thicket of fallen-ness, both within and without. Sin still desires to "reign in your mortal body" (Rom 6:12). Satan inflames that sin by whispering justifications to our already-established and familiar flesh patterns (the closed system this book has been working to describe and map so that opening it to newness becomes possible). Can we not but weep over the stubbornness, the recalcitrance, the sheer cussedness of our fallen

---

1. Lovelace, *Dynamics*, 133–44.

FLESH (ANESTHESIA)

hearts? Indeed, we must weep. Yet, after tears, we must press on in renewed softness—i.e., the softness of having become freshly impressionable by God. Our battle consists of repenting of the flesh and walking out our new life by making Christ-authorized choices in real time. This means we come to the old forks in the road and (more and more often) go a different way. We actually "lay aside the old self" and are "renewed in the spirit of [our] mind" and "put on the new self, which in the likeness of God has been created in righteousness and holiness of the truth" (Eph 4:22–24).

Stage one: we are dead. Stage two: we come alive in Christ; a miracle of newness nestles inside us. Stage three: we battle to strengthen our aliveness in increasing freedom and in loving relationships with others. In the language of this book, we develop an open system—that is, open to God and new life. Consequently, we are more and more free from our old, closed survival system—our old, stale map of how to make life work on our terms.

The new life God wants for us provokes in us both wonder and fear. The wonder is that there really is a miracle of newness inside us. The fear is that there really is a miracle of newness inside us. How can the same reality have both effects? Why are we afraid? For an answer, let's go to an incident in Jesus' journey to the cross. Judas has just betrayed him with a kiss, and Peter, reacting protectively, slices off the ear of one of the arresting party. Jesus responds by saying, "Put your sword back into its place, for all those who take up the sword shall perish by the sword. Or do you think that I cannot appeal to My Father, and He will at once put at My disposal more than twelve legions of angels?" (Matt 26:51–53). Here, Jesus proclaims that power and control, represented by Peter's sword, do not convey the impact of God's kingdom. The kingdom of God does not come with flags flying and tanks rolling, so to speak. Rather, God's kingdom comes in relinquishment of power and in trusting "our Father who art in Heaven" (Matt 6:9). Jesus gives up the power of deploying an entire army of angels and trusts the Father's call for him to journey to the cross. Our fear, then, is that we're called to relinquish our habits of control, called to "put [our] sword back into its place."

CLEANUP

Recently, I was discussing with a counselee how hard it is to face our emotional pain. How it takes great trust in God. How "we walk by faith and not by sight" (2 Cor 5:7). Thinking of a Christian song that had helped her face pain, she said, "Once you jump off the cliff, you either find solid ground or you learn to fly."[2] I said, "Yes, those could happen, but let me offer a third alternative: redemptive splatter. In other words, God doesn't always—though sometimes he does—protect us from hitting the bottom. But God can and will take our splattered pieces and put together a new life." I could see the fear on her face as I said these words. Let me say, I feel the same fear at times. On the other hand, if anyone has the resources to go through pain, it is a believer in Jesus Christ, who knew pain to a searing depth no other has known. How can we know the "man of sorrows," the one "acquainted with grief" (Isa 53:3 ESV), and expect to skate through life "above it all"?

The miracle of newness provokes fear, because we're called to jump off scary cliffs, take the narrow road, lay down our lives. Yet, that same miracle provokes wonder as it dawns on us that we're living in the story of God's having pursued us with an intensity that defeated hell. To defeat hell, he was willing to tear apart the Trinity, in a sense (this idea has to be thought through carefully). To put it another way, we follow a God who loves us so much that he tore open the Trinity to get to us.

Western culture creates a sweet dream where we perform well and are rewarded for our merits by finding "the good life." Self-effort leads to self-satisfaction, or so the dream goes. We end up optimistic (instead of hopeful) in some blithe zone of pillowed happiness. If we're Christians, we get to use God to that end. But happiness is not God's primary concern for us. Charles Spurgeon puts it bluntly:

> I know some of you who are Christian people want to be always coddled and cuddled like weakly babies. You pine for love-visits and delights, and promises sealed home to your heart. You would live on sweetmeats and be wheeled in a spiritual perambulator all the way to

---

2. She cannot remember the name of the song.

FLESH (ANESTHESIA)

heaven; but your heavenly Father is not going to do anything of the sort. He will be with you, but He will try your [heart] and so develop it.[3]

Our flesh, of course, wants nothing to do with this sobering vision. The whole drive of the flesh consists of developing the "spiritual perambulator," the comfort zone that becomes our false god.

---

3. Spurgeon, *Treasury*, 242.

CHAPTER 5

Control and Safety (Sleep)

CONTROL

The agenda of the flesh is to drive for safety and control on *our* terms. Let's start by taking a look at control. The best image I can think of to illustrate control is the scene from *The Wizard of Oz* where Toto pulls back the curtain to expose Professor Marvel working the controls that produce the intimidating spectacle of the "great and powerful Oz." As long as the professor can awe his visitors with flame and thunder and green weirdness, he can bring about the outcomes he wants. "Pay no attention to that man behind the curtain," shouts Marvel into his microphone, still hanging on to the illusion.[1] And this is what we do; we hold on desperately to the effects we bring about through our managed setup of how life is supposed to work. Like Marvel, we hide behind our (we hope) undetectable curtains. Some hide behind pretense, showmanship, and bluster. Some hide within quietness and isolation. Some hide within a nice-guy/nice-girl don't-rock-the-boat persona. Some hide behind a snarky intellectualization. Some hide behind an expertise at

---

1. "The Wizard of Oz," sec. 11.

CONTROL AND SAFETY (SLEEP)

vacuuming out what they want from other lives. Some hide behind a machinelike efficiency, a robotic rage for order. These are broad strokes, but they convey an idea of the flesh's creativity. The reality is that there are as many hide-me curtains as there are people, for the flesh adapts, creating specific responses to new situations that still satisfy our lust for control. Each of us has a characteristic flesh like a fingerprint. Each of us has his or her particular search for control.

In a scene from *The Great Divorce*, C. S. Lewis depicts sin as a red lizard on a man's shoulder. When an angel kills it and throws it to the turf, the lizard revitalizes as a golden shape and grows into a magnificent stallion.[2] Lewis was making the important point that at the heart of sin is a distorted longing. When the sin is destroyed, the longing is revealed as a passion for God, a passion that had been deceived and steered toward a short-term, controllable desire. And that is idolatry in a nutshell. Our desire for control reflects a deeper longing for the defeat of chaos that comes from God as King. But we pervert that desire into quick-fix outcomes that establish us firmly at the control panel.

SAFETY

Safety can be a tricky subject. It is not wrong to be safe per se. In fact, God is a saving God who has pursued us in Christ to bring us salvation, which means deliverance, rescue. So, God wants us safe. Safe from what? From the consequences of our own hostility toward him. The fallen heart, as we've seen, is "dead in [its] trespasses and sins" (Eph 2:1)—that is, dead to God, having no interest in him, because we are deceived willingly into self-sufficiency. "There is none who seeks for God," Paul reiterates in another place (Rom 3:11)—that is, no individual human spontaneously starts a search for God. The fallen human heart seeks its own way as naturally as water seeks the lowest point. Instinctively and without love, the fallen heart reacts to any sense of threat by assessing what "proper" outcomes are missing and then laboring to restore them.

---

2. Lewis, *Great Divorce*, 111–12.

CLEANUP

Does this seem an audacious claim? How can we test it out? Follow me into a story. A boy, age twelve, is playing quarterback for his team as his father, mother, and brother watch from the sidelines. During the game, the boy makes several mistakes. As the mistakes pile up, his father becomes more and more furious. When the family gets home—the father having fumed in silence during the drive—a terrible scene unfolds. The father summons the boy to the front yard and yells, "Tackle me!" The boy protests, but the father will not hear reason. So, the boy hurls himself at his father, who drills him into the ground. This happens two more times. The boy is crying, the mother is screaming, and the father finally stalks into the house. The boy is filled with shame. The mother is filled with disgust. The father is filled with a raging self-righteousness. The younger brother is taking notes on survival. Who is seeking God? There is none.

I can hear my own heart protest, "Can we expect *anyone* in such a scene to be a saint? Who would seek God in that moment?" On the other hand, we could approach the scene through the question "How does the longing of each person reflect a stifled cry for God?" The father longs for significance. The older son longs for love. The mother longs for a husband's strength and wisdom. The younger son longs for a larger perspective to make sense of the chaos. Why don't they cry out to God in the intensity? When a sense of threat floods us, why does that justify our hearts in seeking our own way? Why isn't it easy for us to respond to threat with calm connection to God, who answers our longings as no one else can? Because we have already concluded that God isn't to be trusted. Why? Our "foolish heart [is] darkened" (Rom 1:21). In an unfallen world, it would not *be* saintly for the older son to respond with love and prayer for his father. It would not be saintly, because it would be natural. Why? Because "God is love," and in an unfallen world, the boy would love his father more than loving his own picture of how things should go. Of course, in an unfallen world, the father would not have acted like a brute in the first place, but I'm just making a point here, which is that fallen people are bent toward survival instead of God. Yet, we were created for God, which is why our

CONTROL AND SAFETY (SLEEP)

hearts are restless, as Augustine says.[3] Isn't that the point of Gen 3's story of Satan's tempting Adam and Eve? They were designed to respond to the stress of temptation by worshiping God. The whole pivot of the story is God's disappointed expectation: the very creatures he had designed to love him instead crazily and selfishly turned toward their own resources when they were put to the test. The insanity of sin is that originally, there was no reason for it; and there was every reason for worship. All the humans had to do was worship—to turn to God. Then they (we) would have been fine! All would have been well!

Safety comes from proper worship. As soon as we worship something that is not God, we throw ourselves into danger. For example, when the father worshiped the security and validation that would come from his son's being a whizbang quarterback, the son was immediately placed into emotional and physical danger. The farther we retreat from God, the more we are in danger of forgetting who we really are. The father forgot that his significance comes from God's creating him, that he is already safely identified as God's child if he will but respond in faith. Imagine, then, a secure father on the sidelines watching his son and thinking, "He is making mistakes, but he is trying. He is out there taking the risk. I love him so much! How can I encourage him?" Paul's lament that "there is none who seeks for God" (Rom 3:11) could be paraphrased as "There is none who worships God," which could be further paraphrased as "There is none who seeks true safety." We are safe when we worship.

Deep in what it means to be human is the desire to inhabit, to be safe, to have the constant assault of life halted in repose. On the idea of inhabiting, Gaston Bachelard writes of "the uncommon value of all our images of protected intimacy." He is reflecting on the value of home, of house, of habitation. He goes on to muse that "the primary virtues [of the house are] those that reveal an attachment . . . to our vital space . . . . [W]e take root, day after day, in a corner of the world. For our house is our corner of the world . . . . [It] is our first universe, a real cosmos in every sense of the word." We become aware, he says, of the true value "of inhabited space,

---

3. Augustine, *Confessions*, 3.

CLEANUP

of the non-I that protects the I." He goes on to say that "all really inhabited space bears the essence of the notion of home."[4]

We can't discuss safety, then, without bringing up the idea of home. All creation reels and staggers from losing its true home, from being cast away from its Creator, who once gave it both being and well-being. Now it has its being in God, but its well-being awaits "the revealing of the sons of God" (Rom 8:19). No wonder Paul shows creation waiting eagerly, standing on tiptoe, trembling with desire. Creation has suffered a dislocation that Hopkins unblinkingly describes: "All is in an enormous dark / Drowned." And what is true of "all" becomes still more grim when the poet describes humans: "Nor mark / Is any of him at all so stark / But vastness blurs and time beats level."[5] None of us leaves a deep-enough mark to survive the leveling power of death and entropy. Eventually, every self is "unselved" by the battering of time. All creation groans in response. In fact, it has been groaning with "the pains of childbirth" (Rom 8:22). And where, ideally, should childbirth occur—where should a child be born? Into a home. The image of childbirth is one of homecoming. All creation longs for the great relief of arriving home, of reaching the haven only God can provide.

In a word, then, creation is homesick. Being part of that creation, all people, too, are homesick. There is an aching lostness in the human heart. We belong somewhere, and we are not there. Our lostness fills us with loss. Our fundamental problem? We are made for heaven, and we don't live there. This is our great, inborn pain. I understand "heaven" to be the new creation God will bring about by enacting these words: "Behold, I am making all things new" (Rev 21:5). Ever since being banished from Eden, we have longed for that "garden of delight" (the literal meaning of "garden of Eden"), aching in our souls for wholeness, rest, unmolested peace. I think the first ten thousand years of heaven (if we can talk about "years") will be our saying, "Ahhh!" We'll be relieved of the smoldering, deep-in-the-belly, sour hair ball of stress, the strain of living in a fallen world.

4. Bachelard, *Poetics of Space*, 4–5.

5. Hopkins, "That Nature Is a Heraclitean Fire and of the Comfort of the Resurrection," in *Hopkins Reader*, 80–81 (lines 12–13, 14–16).

CONTROL AND SAFETY (SLEEP)

When I say that "the world is not our home," I mean the world as an organized system of rebellion against God. It is not the same thing to say, "The earth is not our home," for we will, according to the Scriptures, be living on a renovated *earth*, "a new heaven and a new earth" (Rev 21:1). The Bible is not an otherworldly story where God hates matter and wishes us to rid ourselves of our bodies and get away from the tangible stuff of this planet. God gave us bodies, and their being gifts is the chief idea we get from his pronouncing the creation of man and woman "very good." The one who hates matter is actually Satan, since matter reflects God's creative ability and thought. Satan is the one who has torn into God's creation with the goal of darkening and ruining all traces of the divine work. But his is a losing battle, since Jesus has won the cross-shaped victory and has taken captive the things that capture us, an idea based on Eph 4:8. The fact that all things will be restored means that all physical things will be whole, at peace, no longer groaning.

Instead of waiting in faith, however, we're far more inclined to slam ourselves against the gates of Eden, trying desperately to get back inside. Our every effort to make life work on our own terms is that self-effort of returning, of attempting to go home again. But we can't. Foolishly, sinfully, we keep trying and keep right on running into the sword of the cherubim. The Bible's telling of it goes like this: "So [God] drove the man [and woman] out; and at the east of the garden of Eden He stationed the cherubim, and the flaming sword which turned every direction, to guard the way to the tree of life" (Gen 3:24). It doesn't sound safe to try to knock down the gates of Eden. Supernatural guardians are there, and they have an all-repelling sword! We run into all those defenses, and it hurts terribly. Could it be that at times our lives hurt because we are cut by the sword of the cherubim?

Again, it is not wise to hammer on the gates of Eden; but it would be even worse if we got in. Why? The "tree of life" is there, and that is unsafe, because of God's concern that we "take also from the tree of life, and eat, and live forever" (Gen 3:22). That would be a disaster, because it would permanently lock us inside the prison of life on our terms. We'd be forever banned from the story of safety that God is telling. That redemptive story begins in Gen 3:15, which

CLEANUP

describes the already-mentioned battle between righteousness and evil, a long battle that climaxes in the life, death, resurrection, ascension, and coming again of Jesus Christ. Safety no longer lies behind us; we cannot go back to Eden (or any "Eden" we try to build on our own terms). The only safe place is ahead, in God's future, always coming toward us in Christ and prefigured by the outpouring of the Holy Spirit at Pentecost.

Does this mean there can be no safety here on earth? Is all the safety reserved for us in Christ in the new heavens and the new earth? The answer depends on what we mean by "safety." Let's start by remembering that "salvation" and "safety" are related terms. Obviously, a God who comes to save us at such cost is a God passionately interested in safety. Salvation means to be brought to safety, delivered, rescued. But rescued from what? From the horrible outcome of sin—the wretched, blind, endless separation from God known as hell, which is the condition of our realizing forever that our Lord, now no longer available, is all we've ever wanted. This is equivalent to being desperately thirsty beside a sweet stream from which one can never drink. It is to be like the rich man appealing to Abraham in Jesus' parable: "And he cried out and said, 'Father Abraham, have mercy on me, and send Lazarus, that he may dip the tip of his finger in water and cool off my tongue; for I am in agony in this flame'" (Luke 16:24). This is the hell from which we are saved. But we are not only rescued from something dreadful, we are rescued for something splendid. "You have made us for yourself," said Augustine, "and our heart is restless until it rests in you."[6] Imagine that you could take the best food you've ever eaten, the best drink you've quaffed, the best sex you've ever had, the most free and playful moment of your life, an instant of utterly fulfilled longing, the most stunning sunset, the most moving painting, the most touching music, the most stirring conversation—take all that and wrap it up into a gooey sensation ball, and then spread it all over yourself, feeling all those instants in one instant. Being near to God will infinitely outstrip the I'm-blown-away impact of all those rolled-into-one abundances. Our capacities

---

6. Augustine, *Confessions*, 3.

CONTROL AND SAFETY (SLEEP)

to receive will have to be massively expanded to drink from the fire hydrant of God's loving being.

The idea of God's expanding our capacities is also the focus of the Christian's life *before* going home. God doesn't want us to wait to receive the water for which our deepest thirsts are calling. At the same time, God knows we can't handle much inflow in our current condition. The Holy Spirit comes to live inside us—first, to break apart the remnants of the old nature that still cling to us; and, second, to fill us with the passionate joy of the Trinitarian feast/banquet/party/celebration. Neither of these projects is an easy venture, so we walk by faith that God is at work in ways we may have trouble discerning. Sometimes, the Holy Spirit flies below our radar in his determination to break up the hard structures of our fallen-ness. For a penetrating, sobering, hopeful exploration of God's breaking up these structures through adversity, pick up any of Larry Crabb's works, especially his *Shattered Dreams*.[7]

Another way to think about God's breaking up these fallen structures inside us is to imagine them as a false home we have built to make life work on our own terms. Remember the son whose father made the boy tackle him over and over? Earlier, I paraphrased Paul's saying "There is none who seeks for God" (Rom 3:11) as "There is none who seeks to worship" and then as "There is none who seeks true safety." We are bent toward seeking a false safety, building a home in the dark by the best lights we have, which aren't good ones. If we could each have a bright light thrown on our fallen structure, we'd see a rickety shack surrounded by a wall of jumbled stones. But in our dim spiritual vision, we see it as a perfected fortress, adequate, safe, and warm. On another level, since we are made in God's image, we sense that the home is in shambles; but, as we've seen, we "suppress the truth in unrighteousness" (Rom 1:18). We don't see the shambles, we just feel the deep insecurity. We don't read the insecurity properly, so it drives us to pile more stones on the wall.

Our real home will be drastically better. Here is Tolkien's attempt to capture the feel of that home in the Trinitarian safety:

---

7. Crabb, *Shattered Dreams*.

CLEANUP

> Frodo was now safe in the Last Homely House east of the Sea. As Bilbo once described it, that house was "... a perfect house, whether you like food or sleep or storytelling or singing, or just sitting and thinking best, or a pleasant mixture of them all. No evil thing entered that valley." Merely to be there was a cure for weariness, fear and sadness.[8]

Someday, all the wounds of life will be healed; all the shadows on our souls will flee from the light and disappear; every sorrow will lighten up and dance away; all of our sinful responses to pain will melt. For the first time, we will know the delight of loving freely and without self-concern. And the discoveries! Sometimes, I think that God will be like an infinite archipelago of islands, each with an unlimited array of headlands and bays. Each headland and each bay will be suffused with more than enough glory to galvanize the soul to the dance of worship. The child of God rounds the point of an intoxicating headland, and a whole new bay opens up, teeming with its own unique facets of God's beautiful majesty. Belden Lane writes, "Jonathan Edwards suggested that God is far more 'sensuous'—more full of infinite delights, more prone to the endless expansion of relationships, more astonishingly *beautiful*—than anything we can imagine in the stunningly sensuous world around us."[9] That beauty snatches away the wanderer's breath for ten thousand years; and then she starts to explore and, again, worship. Imagine, too, that each act of worship makes the archipelago bigger! That God promises such a home to his children is clear from Jesus' saying, "If it were not so, I would have told you" (John 14:2). We live in the hope of promise—in the hope of a promised home. Now, our home is in hope; then, our home will be God himself.

Earlier, we said that the closed system runs on pain, fear, shame, flesh, control, and safety. Now, we turn to the components that lead to an open system: repentance, moving back through pain, and being embraced by Love.

---

8. Tolkien, *Fellowship of the Ring*, 237.
9. Lane, *Backpacking*, 12; his emphasis.

CHAPTER 6

The Role of Repentance (Waking)

THE ESSENCE OF REPENTANCE

Exploring and worshiping God can, for the believer, begin *now*. But there's a tough reality here that has to be faced. Exploring and worshiping the beautiful Trinity runs aground on our false efforts to build a "home" for ourselves in this fallen world. Letting our true selves grow into being known by God and then knowing God involves an entirely different motivation than that of pounding the world into the safety-and-control shapes we demand. The harder we hammer on life, the more our longing for God becomes obscured, like a stately tree choked by vines. I'm not discussing here our talk about God; we can talk a good game about God, sounding impressively spiritual and "into" spiritual things. Our most common practice vis-à-vis God is, all too often, not worship but the skillful use of God talk that masks at least two things: (1) we want to appear good at this God thing (we construe life as an unending competition); (2) we'd rather manage God than abandon ourselves *to* God. We'd rather show that we're knowledgeable about God than show how inadequate we feel about getting close to him. The truth

CLEANUP

is, the inadequacy is a good thing. Until we admit our inadequacy, we yield to the temptation of being competent in the things of God rather than being dependent on his pursuit of us. Who wants to feel such childlikeness? Not the flesh! We urge ourselves to display competence, not a sense of need.

Our contempt for admitting need blocks repentance. God has *designed* us to need him, but such vulnerability doesn't come easily to us. In repentance, we admit that we don't have what it takes to create a self that both navigates life successfully and does not hurt relationships. We avoid the idea that blazing our own self-protecting trail in the world will entail damaging others. Yet, this is one of the most widespread but hidden practices in our world. If we're going to feel safe and in control, then someone must accept the relational costs. Think back on the father who made his son tackle him in the front yard, shaming the boy for being a "chicken." The demand to have an impressive, confident son on the father's terms eclipsed the son's need to count on his father's unconditional love. The father's demand to feel good about himself brutalized the boy's developing soul. Collateral damage.

Imagine a marathon with its line of runners pounding along. Helpers are stationed at intervals to toss the laboring runners bottles of precious water. Now, reimagine the marathon as the long race of life. Living by the flesh is like the literal marathon in that the flesh taxes the human soul with its sinful energy. The soul is made to be near God, not near the tyranny of the flesh (see Ps 131). Eventually, the human being must off-load the moral and relational costs of living in the flesh. In a reversal of the runners receiving water bottles in the literal marathon, we throw off "toxicity bottles" into the crowd as we run through life. As those in the crowd absorb the toxicity, they sense the cost and burden of the flesh, so they begin off-loading their *own* "toxicity bottles" into those around them. And so it spreads.

Here's an example. A girl grows up in a chaotic home where the alcoholic father has abandoned the family and the mother has moved her children into her aunt's boarding house. They get a room of their own, but the rest of the home is a barrage of traffic, noise, conversation, weird men, and so on. The girl makes a defining

THE ROLE OF REPENTANCE (WAKING)

decision early in life: "I won't feel anything; I can't trust all these people. If I feel, I'll get hurt, so I'm shutting down." Years later, she has her own children, and even in the face of these new lives, she doesn't open her heart. Her children deeply feel her distance from them, each growing up with his/her own pain, sorrowing over the coldness at the heart of their home. Each of them finds life hard to navigate, especially relationally. Each of them off-loads their own hurt in ways that hurt their own marriages. One marriage almost doesn't make it when the wife of that couple crashes into a profound depression. Collateral damage.

But repentance works quite differently than the self-protection in this story. As we run our race through life, we get feedback—let's say, that our impact on others is manipulative, or dominant, or aloof, or some variation on these three. What we're learning at that point is that we've been throwing off those "toxicity bottles." Our impact is not loving but self-interested. We come under conviction; we bow to the judgment that we have more faith in our ability to maneuver through life than in the living God. We see that our God talk has been a screen to hide our relational sin. We have been walking in unbelief and worshiping our own idols (demanded outcomes). We see, too, that our impact on others has often been to take more than to give, to rob more than to heal, to grasp more than to love. We feel a holy fear: "Can there really be a God who is gracious enough to forgive such a one as I?" Then we hear, or read, or remember "being justified as a gift by His grace through the redemption which is in Christ Jesus" (Rom 3:24). We sense both the radical abyss of our sin and the greater downpour of God's grace. We grasp (or are grasped by) God's infinite sufficiency to forgive us based on the work of Jesus Christ. Our defenses collapse, and we are blown by the wind of the Spirit away from safety and control. Blown where? Back through the pain of life, and from there to a godly sorrow, then on toward discovering the reality of love.

CLEANUP

BACK THROUGH THE PAIN

As we repent of our safety-and-control maneuvers, we are more able to see that our relationship with pain affects every other relationship in our lives, for good or ill. In our new and growing freedom from the demandingness of the flesh, we pray differently: "Help me, Lord, not to relate to emotional pain as an emergency. Instead, help me to see that the emergent thing is what you are teaching me *through* the pain." The prayer is that God would transform pain from its old role (spurring us to stay safe on our terms) to its new capacity to teach us, to help realizations emerge (thus, the part of the prayer that says, "Instead, help me to see that the emergent thing is what you are teaching me through the pain").

As we change our view of pain, it can become discovery instead of misery. What do we discover? The main thing is that we've been looking at pain (especially emotional pain) all wrong. We see with more clarity that the flesh has hijacked our pain, turned it into an incessant emergency, and used it to fuel our tenacious drive for safety and control. "Now, we slow down to God's pace," as Larry Crabb says, and see the good in pain—that it's a sign of our homesickness in a fallen world.[1] As I said earlier, "Pain is pregnant with words." Some of these words begin to emerge: "I confess that I've been running from pain because I can't bear the thought of homesickness. It feels too close to homelessness, and I don't want to find out any more about it! The only thing is to run and run and run, hoping to find my own broken-down hut of safety and control. Now, I'm realizing my misplaced faith: I've been trusting my ability to run and to build. I've ignored the fact that I've been loading toxicity onto others, that I've built a shabby hovel, and that I've been faithless toward my Creator." Do you begin to see how full of words pain is? As I said earlier, pain is part of a great conversation, part of a sweeping discourse about suffering that has been developing since the world fell. We reenter this conversation through repentance.

The first place the Spirit blows us, then, is back through the pain of life to arrive at a new understanding of it: pain is not to be fled but consulted. Second, the Spirit blows us toward the "work of

---

1. I owe this wording to a personal conversation with Larry Crabb.

THE ROLE OF REPENTANCE (WAKING)

sorrow."[2] In the "work of sorrow," we wrestle constructively with loss instead of being lost in an agonized sorrow that circles loss but never resolves loss. The secular literature on grieving differentiates these sorrows into mourning vs. melancholy, with mourning being the healthier of the two. While mourning uses a person's emotional resources to accept the permanence of loss, to grieve and to let go, melancholy ties up one's emotional resources in an ambivalent mixture of love and anger. Since the grief isn't clean and simple but has streaks of anger, the grief is tinged with torment, as one punishes oneself for being angry at the lost loved one or lost outcome.

Mourning, or "the [true] work of sorrow," labors intently to accept loss as permanent and irretrievable. It accepts that the lost person, thing, dream, or opportunity is truly gone. It also works to resolve anger at what or who has been lost. The grieving person may, for instance, write a letter to a lost loved one and read it at the graveside. Or, the grieving person may enter a concentrated time of prayer, beseeching the Lord to transform the pain into the hope that God's transcendent narrative will gradually absorb the agony of loss. Mourning accepts the truth that we are not in control of life and that there are things about which we are helpless. It's crucial to bring this pain to speech with the Lord and with trusted others. Again: "Everything must be brought to speech, and everything brought to speech must be *addressed to God*, who is the final reference for all of life."[3]

The idea of bringing pain to speech reflects our nature as people made for story, besotted with story, immersed in a sense of story. A Jewish proverb says that "God made people because he loves stories." While I appreciate the thought, I'd rather say something like this: "When God's Trinitarian story overflowed into creation, God gave the created ones his storied nature." The Latin phrase *post hoc ergo propter hoc* ("after this, therefore because of this") applies here: for humans to be storied folk after creation, it is because of the nature of the creator. One can imagine the Trinitarian conversation: "Since we have danced and told stories for eternity, let us pick one of

---

2. Tolbert, "Voice, Metaphysics, and Community," 148.
3. Brueggemann, *Message of the Psalms*, 52; his emphasis.

CLEANUP

those stories and make it real." The problem of sin looms large here (did the story get away from them?), but that's for another discussion. For now, we can say that a story can't be judged until it is over.

When we bring pain to speech, we inevitably begin telling a story. How did the pain begin? What is at stake? Will the pain be interpreted or ignored? How can the pain be interpreted—what are its messages? Will it be productive or go to waste? Why? Why not? Is there a community that can hear the pain? Why? Why not? Should I/we foist the pain on someone else? All of these questions unfold as answers that form a narrative: "Well, the pain began when my mother emasculated my father for twenty years . . ." How did she do that? "She had this habit of wearing him down with guilt." Why would she do that? "Well, because she grew up under a regime of guilt, because her parents thought they were from the 'wrong side of the tracks' and wanted to prove they were respectable. They found that loading their children with guilt was a great motivator. Having absorbed all that guilt, my mother wrongly loaded it onto others, because she thought it would motivate them to follow her agenda. In fact, there was this time when . . ." And we're in the middle of a story.

While we love stories, and we have a culture of stories, we also live in an age when narrative is played down in favor of facts. The downstream flood of what many have called "the Enlightenment project" has, for all its considerable benefits, burdened the West (and increasingly, the world) with a wall between fact and value, between "what is" and "what ought to be." For example, the fact is that we know how to apply massive quantities of chemical fertilizer to increase farm yields. But *ought* we do that? Science cannot tell us, yet the Enlightenment worldview says that science will discover the facts of the natural world such that our problems become increasingly solvable. So, "is" becomes "ought." An exaggerated faith in science has led to the application of those fertilizers without the caution of a sustained ethical discussion (although that discussion is now unfolding—whether or not it is too late is another question). Note that I'm not saying we should reject science, only that it's necessary that we frame the sciences with an ethics that science cannot itself supply.

THE ROLE OF REPENTANCE (WAKING)

For our purposes, the main point here is that there is a great thrust in our culture toward the gathering of isolated information without a narrative to make sense of it. It is often hard for us to say what our lives mean. If our lives are just "one damned thing after another," they are no more a narrative than is a grocery list (one could argue that there is plenty of narrative behind a grocery list, but that's just the point: the story is behind the list; the list itself is just an itemization). If life is just one day after another at the same stultifying job and night after night watching soulless television or video streaming or whatever other screen hypnosis one might prefer, then life is not a story. Often, we are simply watching the stories of others on these screens, but not entering a story of our own. Part of our hesitation is that stories are full of conflict, and one of our core values is to save face by avoiding conflict. Ergo, to be healthier, we need ways to rehabilitate conflict.

So, let me tell you a story. You may have picked up from reading this book so far that I am a counselor by profession (and a spiritual director by trade, you might say). The upshot of this is that I am constantly in the middle of stories. Here is one of them: I am working with a person who has three tendencies, which are to act as an adult, to act as an adolescent, and to act as a five-year-old (or so). The precipitating events that caused her deep internal splits lay in years of sexual abuse in her childhood. The perpetrator forced many sexual touches, all accompanied by a stream of propaganda that he was doing her a favor, preparing her for womanhood, creating a special, privileged secret between them, and other heavy manipulations of reality that played with her mind such that the mental abuse was almost as bad as the sexual. As we've worked together, we've discovered that her adolescent aspect has the job of containing anger and pain. It's as if a bomb went off in this human life, and the adolescent is ground zero. She absorbs the pain and rage while the adult personality carries out the life of a wife, mother, and employee.

The ways the adolescent keeps the pain managed is to blame herself for "participating" in the abuse and to reserve the right to commit suicide. This "right" gives her a sense of control. In our last session, which focused on the marriage (so her husband was

CLEANUP

present), the adolescent asked the husband to leave the room, which he did. She then told me that she had decided that if she were going to harm herself, it would be on a Saturday night (this was a Wednesday). I was curious as to why a Saturday. After some back and forth about this, the discussion morphed into the question of deceiving the husband. So, how do you allow for this voice to speak, given that she's never had a voice? How do you do that in a way that frees her from all the coerciveness with which she was hammered for those abusive years, yet do so also in a way that honors counseling ethics, honors her husband and children, and protects the whole person from suicide? I felt tremendous pressure in that session. The husband came back into the room after ten minutes or so, and his wife simply would not talk. I didn't feel the freedom to betray her secret in view of my concern to keep trust and avoid all coerciveness, yet I felt the husband could not continue being kept in the dark. How could I risk that he end up with no wife and with motherless children? Yet, I could not find the adult aspect of the woman, could not get her back into the session, so powerful was her internal conflict. I'd lost an ally.

The session ended owing to time constraints, and I resolved to call the next day. I got in touch with her, and we basically had a conversation that entailed my talking to a rapid oscillation of the conflicted voices inside her. We debated over self-harm, and she made a promise that she would be present at our next session. That night, I got an email from her saying that she wanted to tell her husband the full truth, but a previous counselor had told her that suicidal ideas should only be told to a professional counselor, not to a husband. I wrote back that I disagreed, saying that God had designed marriage to develop an intimacy and sharing that was keyed to the depth of intimacy in the Trinity. In other words, she should open her heart to her husband (note: the husband is a good man; there are some husband-wife situations where I would not have given this same response). Her response was priceless:

> I received your email and I just couldn't wait until _____ got home to talk to him so we texted a bit. I told him about the _____, and when I am in a bad place I can only see two options, either taking _____ or _____.

THE ROLE OF REPENTANCE (WAKING)

> For some reason the first option seems like a better option because (at the time) it feels less violent. I also told him there was some kind of plan, and that I was frightened about it. I asked him if I tell him my thoughts would it mean we were together and that he cared about me and that he would also be physically beside me. I would like to quote what he sent back because I absolutely love his response. "Yes we would be together! Always! And both of us together in a troubled place is better than one of us alone in a troubled place." After that I let him call me. . . . I was feeling very protected, loved, and not alone.

When I read her email, I began to weep with joy, with relief, with sorrow. Joy, because God so clearly revealed that he is telling a good story in these lives. Relief, because of the deep conflicts in my heart about walking a tightrope between relating to her as someone having personhood instead of being a mere thing, and doing this while somehow protecting the whole person—body and soul. I felt sorrow, because the whole reason for all this wrestling is the selfishness of the man who abused her, the man who failed to know his own story well enough to learn from it rather than acting it out. He didn't throw out toxicity bottles, but rather barrels.

The idea of conflict and its rehabilitation is what set me thinking about this story. On one level, I wanted no part of this conflict. I wanted no part of the war with myself over the wise way to proceed, along with the conflict with the woman over her willfulness. Yet, if I had escaped the conflict, I would have had to exit the whole story. If I avoid enough conflicts and therefore enough stories, eventually, I have an empty life. Revaluing conflict is crucial. While there was plenty of conflict and stress with this story, there was no other way to experience the blast of joy that came with a wife breaking through an enormous inner wall to trust her husband and become the beloved woman she'd always wanted to be.

CHAPTER 7

Love (Reaching)

We've seen that the Holy Spirit longs to blow us away from safety and control on our terms. From the "safe harbor" of our defenses, the Spirit longs to blow us along with strong gusts of repentance so that we journey back through the pain. The goal is to transform pain from a driver to a teacher. The Spirit's longing is for pain's real messages to come to light, since it is going to be there anyway. It might help to think about it this way: the word "angel" means "messenger," so why not begin to ask ourselves, "What is the 'angel' in my pain? What is the messenger seeking to say?" Why not have pain become what it really is—an announcement that we are made for a better world, that we are made to come home? And when I speak of "a better world," I don't mean an eternity of sitting on some cloud like a heavenly bleacher bum. I mean the world that explodes onto the scene through the scrap-pile aftermath of God's rolling up the universe like an old shirt (Heb 1:12) as he creates "a new heaven and a new earth" (Rev 21:1). We're no more privy to how God does this than we are to how God originally spoke all things into being in the beginning. The point is that we will be embodied people in a material world covered in God's glory, so that all things will open up their real song and "clap their hands" in joy (Isa 55:12).

Why will this fantastic scenario happen? Because God loves all that he has made. God is fighting for it to return to its real, whole

LOVE (REACHING)

shiningness, working always to restore us and our world. God is the Creator and is delighted with what he has made. The Lord loathes the evil one's pretentious efforts to haul God's masterpiece off to the dump. God hasn't stood for it, won't stand for it, and will continue to stand against the evil one's lying pride and waspish malice.

When I speak of the Spirit's blowing us back through the pain to love, I see love from three angles. First, I see God's own love for us: "I have loved you with an everlasting love" (Jer 31:3). An everlasting love is an outlasting love. It will outlast anything that shatters our hopes and dreams in this world. A love like that expands the focus of our lives. If the love of God will outlast sickness, loss, abuse, weakness, abandonment, shame, rejection, foolishness, evil, and so on, I can lift my eyes from the limited horizon of my self-concern. I can ask, "What is God up to around me? What are God's kingdom concerns, and how can I be a kingdom citizen who loves as he loves?"

How much does God love us? Inexpressibly much. Here's a true story designed to shed light on that matchless love for us: A fire broke out in Mann Gulch, Montana, in 1949. Fifteen smoke jumpers parachuted in to fight the fire. They gathered on the side of the gulch opposite the fire, making sure they had an escape route to the nearby Missouri River. But the wind changed, and the fire jumped the gulch and cut the firefighters off from the escape. Worse, the wind was blowing the fire up their side of the gulch right toward them. The men ran toward the ridgeline, hoping for safety on the other side. But the crew's boss, Wagner Dodge, saw that the fire would overtake them before they could get over the ridge. In a moment of instinct, he simply lit a fire in front of him. As it burned away from him, he stepped into the growing, burned-over area, now fuel-less. He shouted and beckoned for the others to step in with him, but they thought he was crazy and preferred to trust their legs. They kept hoofing it for the ridge. The crew chief knelt face-down in the ashes and survived. Of the fifteen who headed for the ridge, only two made it.[1] Wagner Dodge's inspiration to light the escape fire was brilliant. But imagine: What if he'd had to light *himself*

---

1. MacLean, *Young Men and Fire*, 92–109.

CLEANUP

on fire to get the escape fire going? This is what Christ did, giving himself into the fire of the crucifixion so that we could step into safety from the fire of judgment. And from the fires of fear, shame, hopelessness—indeed, for all the things that burn us. "Come here!" he calls. "Come to the safety in which I burn, creating shelter for you. And I am still burning for you! Come!" Were we to ask, "Why are you still burning, Lord?" I believe he would say, "The love I have for you is burning me. A fiery passion for your well-being lives in my bones, and I will not be well until all my children are home." We have it this way in Scripture: "Greater love has no one than this, that one lay down his life for his friends" (John 15:13). We are his beloved.

The second aspect of love I want to develop is our love for God. First was his love for us; now, our love for God. Let's say my love for God is a tall glass of water. I love God by pouring out this water in obedience, in worship, in ministry, in caring conversations with others, in going the extra mile, in service, in witness. In fact, whatever I do as unto God can be seen as my pouring out this water to him. Now, picture that I can section off this water into two parts: loving God out of duty vs. loving God because my heart is full of gratitude to the marvelous Giver. For much of my Christian life, I'd say that 80 percent of that water was that of loving God dutifully, because a good person would. Twenty percent was truly heartfelt and amazed. Do you hear the problem? The phrase "because a good person would" betrays me out into the open as one who was trying hard to be a good person. There is something arrogant about that position: I'm working on being a good person, and one way I show my progress is by loving God. The problem there is that I'm using God to demonstrate my progress toward being good! I need a "good" version of myself that I can use to prop up my self-concept. This amounts to using God as a tool, and it reminds me of the disciples arguing among themselves about who was greatest in the kingdom of heaven (Matt 18:1). Jesus simply puts a child among them and says, "Unless you are converted and become like children, you shall not enter the kingdom of heaven" (18:3). We so easily put the cart before the horse, wanting to validate ourselves by using God ideas, God talk, and God activities.

LOVE (REACHING)

"We love, because He first loved us," says John in his first letter (1 John 4:19). Only as we see the size of God's love (infinite) and the passion of God's love (giving up his own Son, who would set himself on fire to give us a place to stand) do we grasp what it is to love him back. We love because Jesus crashed onto this fallen planet just as we crashed so that he could un-crash us. The thought behind the crash imagery is that to be born into a fallen world is already a collision for us. As we learned in previous pages, our fundamental problem is that we're like fish washed up on the beach. Have you seen them? They are *not* comfortable! They don't set up little beach chairs and open a bottled water for some refreshment as they call for an umbrella and some summer reading. Instead, they flop about in an agonizing search for water, for the cool world out of which they've just been hurled. A human born on this planet has the same deep-down sense of being hurled into desolation.

But! We have not been abandoned in the wasteland! In the birth of Jesus, God became human in a radical act of solidarity, saying, in effect, "I am with you; Immanuel—'God with us'—is my name!" Christ crashed onto our planet in order to clash with the devil, death, sin, demons, sorrow, foolishness, darkness. Why do we love God in return? Because he fought for us and still pursues and intercedes for us. Because right now, at the right hand of his Father, Jesus still inhabits an earthly body. What a sign of the limits Christ entered into for us! What an evidence of sacrificial love! The second person of the Trinity forever bearing the resurrected body of beloved humanity. If ever we wonder whether we're loved, we can see it in Jesus' having kept our own form. Out of sheer love, he has agreed to remain one person with two natures forever. And in his human nature, he has a human body. Forever. The message to us is "I will not be without you."

The incarnation, then, sheds new light on the command "You shall love the Lord your God with all your heart, and with all your soul, and with all your strength, and with all your mind" (Luke 10:27, quoting Deut 6:5). The sheer size of God's heart for us tugs our hearts to respond in love. In other words, Jesus put himself in danger to get us out of danger, risking and giving his life that we could be set free from risk. Former castaways, we are now warm and safe on a ship bound for home. We can reflect back on hopelessness, then (with its gasp of anticipation) on having seen sails at the horizon, and then being picked up by boats and ferried to the ship, receiving clean clothes and our first good food and drink in forever. And how did we get marooned in the first place? We'd mutinied against the same ship's company that now sees to our every need and takes us home! How we are indebted to grace!

What does it mean to love God in return? First, it means to intensify that passion for him that begins when we're reborn in Christ. Second, it means to worship. Third, it means to follow. Fourth, it means to love as God loves.

First, to strengthen that passion for God that began when we were reborn: "If any man is in Christ, he is a new creature" (2 Cor 5:17). God gives us a new heart as part of our new birth, a heart that is softened toward him and now longs to "know Him, and the power of His resurrection and the fellowship of His sufferings" (Phil 3:10). To know God requires that we spend time in solitude and silence, learning to recognize God's movements within our new hearts.

On the other hand, knowing the Lord means war. Why? Because we are not yet free from the flesh, which, as we've discussed, is the sneak-thief inside that still whispers, "Follow your own way apart from God." I reiterate: "The flesh sets its desire against the Spirit, and the Spirit against the flesh" (Gal 5:17), so that the Christian life means that we wrestle against ourselves. But it is not a dualistic wrestling with no certainty of outcome. Rather, it's a wrestling in the present but accompanied by the promissory note of Christ's finished work that has been applied to us by our faith in him. We wrestle in hope. Brunner puts it well:

> But the whole man is not faith; faith struggles to free itself from unbelief, from sin, it strives to wring union with God out of the contradiction, the new nature out of the old nature, the "Yes" out of the "No." The actuality of faith is the new man [in Christ]; yet the eggshells of the old nature still cling to him as something which has been overcome, but still also as something which has to be overcome again and again. The old nature ... made itself substantial as "flesh"; its movement was arrested; in the

LOVE (REACHING)

act of tearing itself away [from God] it became rigidly fixed as habitus [a persistent disposition] . . . . The decision, however, that has been made in Christ is total and universal. It is a whole reconciliation . . . but the decision by means of which we appropriate that Christ-decision in faith, by which as it were, we take it over, is not in this sense total, not definitive, but it is a process which is still going on, and is not yet completed. . . . [It is] a continual "I have said yes," which lasts as long as life itself.[2]

Brunner's picture of the "eggshells of the old nature" that "still cling . . . as something which has been overcome, but still . . . has to be overcome again and again" captures clearly the war inside the Christian. To know God through Christ by the Spirit, then, is to know contradiction. Walking with God throws us into a battle with ourselves, the world, the flesh, and the devil. "Contradiction" literally means "saying against," and we are constantly intruded on by sayings against God and against our true selves in Christ. We must respond with passion—with a deep, surging concern—lest the lies against God and against our true selves infiltrate us. We must push back fiercely with truth, engrossed in the campaign to liberate acres of the heart over which Satan has claimed squatter's rights. He has released a toxic spume of chill gloom, a Mordor of the soul, a hectoring trespass. No believer can accept such violation as a *fait accompli*. Such would be tantamount to surrendering someone to be raped. We must fight twisted passion with the honest passion of the redeemed self, grounded in the suffering hope of Christ.

This brings us to the second of the four ways to love God in return: to worship him. To worship God is to acknowledge and express his worth. As Everett Harrison observes, "Our English word means 'worthship,' denoting the worthiness of an individual to receive special honor in accordance with that worth."[3] Worship, boiled down to its essence, is "pure adoration, the lifting up of the redeemed spirit toward God in contemplation of his holy perfection."[4] Jesus says, "The true worshipers shall worship the Father in spirit and

---

2. Brunner, *Man in Revolt*, 488.
3. Harrison, "Worship," 560.
4. Harrison, "Worship," 560.

truth; for such people the Father seeks to be His worshipers. God is spirit; and those who worship Him must worship in spirit and truth" (John 4:23b–24). My own definition of worship? Lifting up to God a passionate gratitude, because he has come through the devil, death, hell, demons, fear, defeat, sin, sorrow, and despair—all this to rescue us at great cost to himself. When I say that God came through all these things, I mean the one God who acts in a unity of three Persons: Father, Son, and Holy Spirit. So, above, when Jesus says, "The true worshipers shall worship the Father in spirit and truth," I think it would be better to capitalize "Spirit and Truth," the first word referring to the Holy Spirit and the second to Christ, the Word—i.e., the Truth (he self-identifies as "the truth" in John 14:6, the familiar "I am the way, the truth, and the life" passage). Jesus' description of worship in John 4:24, then, is Trinitarian: "The true worshipers shall worship the Father in alliance with his Spirit and Christ Jesus" (paraphrase). When I said a few lines ago that "he has come through the devil, death," etc., I keep in mind that the whole Trinitarian interplay of love has been mobilized in the one God's declaring a massive no to the insult that Satan has hurled at God's work—i.e., the great splotch of black so ruinously applied to God's beautiful creation. In worship, we speak and sing our agreement with God. We join together as the people of God to declare a passionate yes to the faithful, Trinitarian no to sin that is God's light in the darkness.

The third way we love God in return is to follow him. This sounds pretty simple, right? Following means doing what Jesus does: "If you love Me, you will keep My commandments" (John 14:15). Not a hard concept, yet also super difficult to pull off. Listen to Jesus tell about a central theme of his journey: "Uunless a grain of wheat falls into the earth and dies, it remains by itself alone; but if it dies, it bears much fruit" (John 12:24). He unpacks this more: "He who loves his life loses it, and he who hates his life in this world shall keep it to life eternal. If any one serves Me, let him follow Me; and where I am, there shall My servant also be; if any one serves Me, the Father will honor him" (12:25–26). Following Jesus means giving up the me-first, grasping-for-life energy our world constantly promotes. We go where Jesus goes: into loving sacrifice based in

LOVE (REACHING)

deep trust of God. As we've discussed throughout this book, following Jesus entails a massive shift in the master values of our lives. It's a radical challenge to let go of a picture of life in which we power up to have leverage over events, circumstances, and people.

This line of thought carries us to the fourth way we love God in return: by loving as God loves. What does it mean to love well, to love in God's manner of loving? It means that we abjure the "What about me?" energy that too often lurks in our hearts. Urgent, insecure, and querulous, we ask, "Am I losing ground? Falling behind? About to be downgraded like a doubtful draft pick? Close to being voted off the island?" These questions reflect something we talked about earlier—Luther's idea of self-concern doubling back ever more to check on itself. The self-focus of "me regarding me" simply strangles love, because love straightens up to God and reaches out to others. Love uncoils from self-absorption, stretching out toward the concerns of God and of the other person. Love wants to connect with the story God is telling about the other person. Love wants to help the other disconnect from the false stories that entice all of us when we are turned inward to ourselves. "Love one another" (John 13:34) also means that we can't remain indifferent to the false stories that wind around other souls like the ivy that chokes a great tree. We look for openings that allow us to engage those wrong narratives. And we hope others will love us enough to inquire into the wrong stories we are telling ourselves.

What makes these stories wrong is that they are influenced by the lies that God can be replaced and that it's up to us to replace him with our own resourceful ingenuity. Again, these are operations of the flesh. As soon as we buy into these two ideas (to replace God and to do so with your own ingenious schemes), a whole world of false stories intrudes on us. The lies embedded in the stories stifle love, for love is based on truth.

For example, Bill was a guy born into a family where multiple generations had handed unresolved issues down the generational ladder, deeply polluting the relational environment. The noise generated by family members as they contended for limited emotional resources made Bill feel unheard as a child. It took so much to get on his parent's "radar screen" that he finally concluded it wasn't

CLEANUP

worth the battle. But the pain of that defeat introduced him to a lie: "You don't have a voice." The lie festered inside, filling him with a pervasive ache. The lie also gave him something to work on—that is, to prove that he was worth hearing, that he did have a voice. Bill had developed a false story: "Once upon a time, there was a boy who had no voice, so he knew he had to strive by might and main to unleash his voice."

Years later, Bill found himself arguing with his wife, Darla, about how to load the dishwasher. The backstory is that he had lost his job, and it took him eight months to find a new one. The problem was that the new job was forty-five minutes away, and he had to be there at 6:30 a.m. Not being a morning person, Bill proposed an idea to his wife: "Load the sharp knives downward in the dishwasher, and the blunt knives pointing up. That way, I can just grab a good knife to butter my toast on my way out the door. I'm going to be in a hurry." Why all this? Because the couple's habit was to load their dishwasher at night and then get their utensils and crockery out of it in the morning. Darla agreed to the plan.

At first, things went well; they took turns and loaded the dishwasher as planned. Bill had easy access to that perfect knife for several mornings in a row. He was even getting to work on time. Then Darla started to neglect her side of the agreement. When it was her turn to load the dishwasher, she would sometimes forget and hurriedly throw the utensils every which way into the loading basket. Bill felt frustrated, but he didn't say anything. Darla forgot more and more often until Bill got up late one day, was in a bad mood, and felt pressure to get out the door only to have to search for the blunt knife in what seemed like a forest of sharp blades. He snapped at Darla, "What is this, rocket science?! Sharp knives down and blunt knives up! How hard is that?" His wife didn't take kindly to such rough handling, and she shot back, "You and your stupid plan! How hard is it to sort through a few knives?!" For five minutes, they had a verbal firefight, and then they had to rush off to work.

If you had been there, you'd maybe have thought something like "Here are two grown adults fighting about how to load the dishwasher! How pitiful!" But the fractured plan for how to load the dishwasher was only the tip of the iceberg. Beneath the waterline,

LOVE (REACHING)

so to speak, was the old story. When Darla forgot to load the utensils as planned, Bill felt the old pain in his heart: "You have no voice." His instinct was to unleash the suppressed voice, an urge he struggled to repress. But one morning when he was late to work and the pressure was on, the old story came to a boil, and he did unleash that voice of his. In the flesh, he believed the old story and believed the old solution. Where is love here? Gone. Is Bill loving as God loves? Clearly not.

Like ivy choking a great tree, as we saw earlier, the flesh chokes out love. That image of the smothered tree brings us back to our beginning point of using environmental and ecological ideas to describe how the flesh damages relationships. We said that human flesh patterns can be viewed from an ecological standpoint. The flesh damages the "ecology" of relationships, because it flattens them into a stiff exchange of defenses in which people are armored up. Why? Because the flesh is whispering, "Watch out for threat! You could get hurt! Be ready to mobilize! No one is there for you, including God! Be vigilant!" Much of the time, we could drop the exclamation marks and say that the flesh is a whisper. That's because this program runs mostly in the background, well below our immediate awareness. Why? First, because if it "ran" on full alert all the time, it would quickly drain our reserves and leave us exhausted. Second, we are so skilled at our defenses that we feel protected most of the time. Our shields are up and our strategies are working, so our monitoring of threat runs at a low level of alert. This effortless ability of the flesh is an example of what one researcher calls "skilled incompetence."[5] It's skilled, because it runs without our having to pay attention, like a jetliner on autopilot. It's incompetent, because it proceeds in ways that are bad for relationships. To put it differently, what the flesh does well (hiding our me-first motives) actually means that relationships will suffer. This is true, because relationships work from the nutrients of love, vulnerability, self-sacrifice, hard work, generosity, mutuality, kindness, and undefended honesty. The flesh is as allergic to these things as it is to God. Why?

---

5. Argyris, *Organizational Defenses*, 21.

CLEANUP

The Bible tells us that fallen humans interpret the nutrients of relationships (love, vulnerability, etc.) as weakening our defenses. As indeed they do. How could it be otherwise? God is love, and believers are called to love as Christ loves. That simply means loving others with fewer and fewer defenses. That means a journey toward risk. It means making a disciple's journey from being a closed to an open system.

When the flesh is the engine driving us and our defenses are threatened, we feel like an exhausted squad of soldiers in a breached citadel: here comes the worst—here comes annihilation. It feels like being thrown into a bottomless pit. We fight against this possibility with all we've got. I remember asking one man, "Why do you react so angrily when your wife wants more from you?" He said, "I feel that there's an abyss behind me, and she's pushing me toward it. I have to push back as hard as I can, or I'll fall in." When I asked him what would happen if he fell in, he just looked at me with a shudder. "Dread" was the only word he could say. Over the years, I've heard similar thoughts from many men and women. I've felt this dread in my own life. I've watched others avoid this same dread through all kinds of urgent maneuvers. Why is this dread so common? Why is it at the roots of our existence? Our culture tends to answer that such dread is adaptive, that it has some kind of survival value. Or the answer can tend toward a common-sense response: "Who wants to look stupid or be taken advantage of? Of course, we're going to push back. It's just logical."

But such answers are superficial. The human heart feels this dread not because of some justified defensiveness, as if we're innocent folks just trying to survive. The real problem is clarified in the early chapters of the Bible. There, Adam and Eve initially stand before one another naked and unashamed, as we've seen (Gen 2:25). When they sinned, they covered themselves, a response to shame. But why the immediate instinct toward self-protection? What did they now know that made shame such a dreaded outcome? The answer is that they now knew that there was an abyss waiting for them. God had said clearly, "The day you eat of [the tree of the knowledge of good and evil], you will surely die" (Gen 2:17). The abyss they dreaded was this new thing called death. Let's return to

LOVE (REACHING)

the man who pushed back when his wife wanted more. His feeling of being at an abyss was not a dread of something she could do to him, although that is what he felt in the moment. The truth is that he was making a category mistake. He categorized his wife as being able to destroy him, but what's really true is that we, as sinners, dread God's power to let us go into the death our sin has brought on us. We deceive ourselves into thinking it is people of whom we are afraid.

Jesus corrects us by saying, "And do not fear those who kill the body, but are unable to kill the soul; but rather fear Him who is able to destroy both soul and body in hell" (Matt 10:28). Hebrews 2:14–15 gets at it this way: "Therefore, since the children share in flesh and blood, He Himself likewise also partook of the same, that through death He might render powerless him who had the power of death, that is, the devil, and might free those who through fear of death were subject to slavery all their lives." "Children" here means human beings—all of us. "He" is Jesus, who "partook of the same"—that is, he took on our own existence, becoming human, just as we are. And this led Jesus to death, but not just any death—death on the cross. Since death itself could not hold Jesus, death died in the sense that it can no longer hold those who believe in Christ. According to Heb 2:14–15, Jesus' gift rendered "powerless him who had the power of death, that is, the devil." Our sin has not only brought death on us but has also given the evil one the right to trespass on our lives and impose death. The original horror story. But Christ has freed us from this nightmare, and that freedom becomes ours when we believe in him. This gift is captured in the words "and might free those who through fear of death were subject to slavery all their lives" (Heb 2:15). The man whose wife wants more of him dreads the abyss not because his wife really has the power to push him into some deep hole of nothingness but because he is caught in his slavery to the fear of death. He transfers his fear to her, because he deems he has some chance of winning. The real battle against "him who has the power of death, that is, the devil" (Heb 2:14)—that one he has no chance of winning.

The Psalms often use the image of the "pit" to convey this fear. For example, Ps 69:15 says, "Do not let the flood overtake me,

CLEANUP

neither let the deep swallow me up, / And do not let the pit shut its mouth on me." Whether the flood catches me or the deep swallows me or the pit chews me up, the effect is the same: I'm a goner, and I have no defense. This complete helplessness is what the man fears when his wife wants more. The nerve she hits is his primordial fear of eternal death. The solution to this fear, in our typical operating style, is to deploy a set of "justified" defenses. But this is not the Lord's way. Instead, God calls us to an exercise of memory. Through purposeful remembering, we retrieve something precious: the story of who God really is, what Jesus has done, and how Spirit-led purposes apply to us. In other words, the Spirit calls us to the practice of retelling our stories such that we are recalled to love and thus to real living instead of our practicing the arts of defensiveness.

I want to use Paul's idea of the in-Christ relationship to deepen our understanding of God's call to love. As we have seen, the apostle says in 2 Cor 5:17, "If any man is in Christ, he is a new creature; the old things passed away; behold, new things have come." The Christian, then, is placed into Christ by his/her faith by grace (Eph 2:8-9). Faith involves a repentance from blindness (that of being asleep, obtuse to the God who is self-revealing). Now seeing, the believer turns (this is conversion) and is placed into Christ (this is adoption). This sphere of new being releases the believer into the freedom of a new becoming. That is, one had previously become a person marked by a lesser life, the life of someone "dead in [their] trespasses and sins," one of "the sons of disobedience," indulging "the desires of the flesh and of the mind" (Eph 2:1-2). This person was living in "the futility of their mind" and was "darkened in their understanding," marked by "the hardness of their heart . . . , callous," given over "to sensuality, for the practice of every kind of impurity with greediness" (4:17-19). This journey of a dark, desiccated, sinful becoming—this decline and deterioration—is now invaded by the new creation, the "light of the knowledge of the glory of God in the face of Christ" (2 Cor 4:6). Now, the new, in-Christ reality dawns as sheer gift.

The grace of newness, of being in Christ, doesn't just park itself in the believer's life. It brings with it a new stance, a move away

LOVE (REACHING)

from "the hell of self and self-consciousness,"[6] a move to become a person marked by self-giving, first to God and then to others (again, "And you shall love the Lord your God with all your heart, and with all your soul, and with all your strength, and with all our mind; and your neighbor as yourself" [Luke 10:27] and, "We love because He first loved us" [1 John 4:19]). Love from God becomes love back to God, which becomes love of neighbor. The first component—love from God—is all-important and primary. It is so because it's the love that flows from the perichoresis (the constant, loving interchange) among Father, Son, and Holy Spirit. This Trinitarian love expresses itself through self-giving that somehow doesn't consume the self. The Father remains the Father, etc. The self-giving interacts with self-consistency, yet a self-consistency not of selfishness but of standing by one's word, of being willing and able to be counted on.[7] In the stance of "I commit myself" arises a self free from self-regard, self-anxiety, and self-aggrandizement. This free self opens its hands to love well, not calculatingly or self-interestedly. Here, then, is my (inherently lacking) effort to describe the love that flows within the Trinity.

This Trinitarian feast of love exerts the pressure of a story (as all stories bring to bear some type of pressure or influence). The Trinitarian story presupposes the incarnation—the climax and verification of God's self-giving—and reaches out in Pentecost, when the Spirit comes as gift to found the church and to "convict the world concerning sin, and righteousness, and judgment" (John 16:8). God isn't a hidden God but a pursuing, loving God whose mission is to make "all things new" (Rev 21:5). One of the pressures within this story is that of conviction.

Why does the Holy Spirit "convict the world"? Just as there's a built-in, festive love within the Trinity, there's a built-in, fallen hatred within the world: "For everyone who does evil hates the light, and does not come into the light, lest his deeds should be exposed" (John 3:20). Evil is the result of sin, with its compound of pride, anxiety, and sensuality. Wrapped in this unholy trinity, sin acts as a

---

6. Payne, *Healing Presence*, 53.
7. Ricoeur, *Oneself as Another*, 118.

CLEANUP

closed system of darkened understanding and self-interest, a drive for personal versions of safety and control that results in a will to power. Regarding the will to power, it's simply a fact that power sees any alien power as a threat to be resisted. The will to power is a will to resist. This resistance, emerging from sin, is culpable. It's culpable, because it resists God's good will to love and because it generates chaos. Creation is anti-chaos; the new creation is anti-chaos. But sin feeds on chaos, even as it generates it. The new creation encounters the sin/chaos compound, judging it, sentencing it, putting it under the divine no. God's strong negation of sin pushes into a fallen world with convicting power, seeking to create spaces for Trinitarian love to invade and move believers to feast on this love that brings love for others.

CHAPTER 8

Synthesis

Relationships call for love. The flesh aims for survival. The two energies are incompatible. To the degree we "are living according to the flesh" (Rom 8:13), we live in the self-interest that quenches love. We become thirsty for love. Searching and searching for it, our souls become deserts. This very dryness can become a gift if we allow it to speak to us, saying, "The water you seek is not available in this world." All too often, we throw away the gift, seduced by the flesh's whiny, tyrannical insistence that we can drill extensively enough to find good water for our dry souls. "Drill" is not a very relational word. Have you felt the drill of others? In your honest moments, can you see how you've drilled into others?

In the emergency of a fallen world, the flesh hoodwinks us into acting unlovingly as we drill for water. Living in the dust, we create more dust. We can't get out of this cycle through our own resources. As we've seen, our only hope is to be shattered into repentance, a radical turn that can only be fostered by God. Our condition can be pictured this way:

CLEANUP

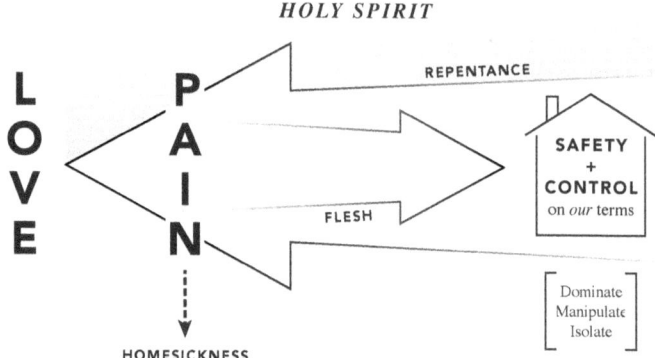

Here, we can see many of the elements of our journey so far. The smaller arrow (pointing to the right) represents the closed system of the flesh with its urgent energy. Toward the left, the word "pain" captures the origin of our dilemma—that is, we're born into a fallen world for which we're not designed but in which we must actually live. In the world of fallen-ness, we must actually deal with the pinching pain of homesickness.

To understand "homesickness," let's return to what we learned about coral reefs. When too many toxins infiltrate the water, or if the water becomes too warm, or both, the very suffering of the coral is a sign that it's "homesick" for the old conditions. So it is with us. Born into adverse conditions, we feel (and try not to feel) the pressure, the very "against-ness" of these conditions, as "the thief comes only to steal, and kill, and destroy" (John 10:10). The discord and drag of a fallen world drowns and scalds our hearts by turns as we yearn for home. This longing is multifaceted. We long for love, significance, security, meaning, hope, joy, freedom. And we live in a world that can only advertise these things.

Thus, we live in a perfect Madison Avenue of claims that the good life is here or just around the corner. The magic door will open if we can just find the right key. If we pursue our bucket list, stay upbeat, develop the right investment portfolio, go to the right university, vacation at the right resorts, grind away at the right job, buy the right jewelry, follow our dreams, believe in ourselves, burn enough calories, go gluten free, locate ourselves on some psychological test,

SYNTHESIS

read enough self-help materials, and so on, we can imagine ourselves drawing ever nearer to the magic door. On the other side of that door, we'll find the answer to our longings, the quenching of our thirst, a sense of home.

But these are closed-universe answers. As such, they're too confining. They're tantamount to promising an orca a good life in a swimming pool.

What I'm calling the "swimming pool" is, in our diagram, that little house of safety and control. While it's small in the diagram, it's huge and terrible in its effects. The "house" represents a lie that the flesh proposes. It's a delicious lie, wheedling its way into our hearts so temptingly because it avers, as we've seen, that life in a fallen world is manageable. As we learn to fashion our own quick fixes of safety and control—so says the lie—life becomes something we're able to master. Those quick-fix tactics collect into a stable pattern, a set of "go-to" options for survival. As they settle into our lives, becoming habits, the flesh offers another lie: "This pattern is your home. You need never be homesick at all. You have all the shelter you need." We become experts at technique, master tacticians. In turn, we come to *need* these survival techniques, and they become our masters. They become the lords of our lives. In serving them, we feel safe. Yet, they're far too restricting, and they mitigate against love. They quench the Holy Spirit and are idolatrous.

In describing these tactics as "huge and terrible" in their effects, I may sound dramatic. Consider the following: Imagine a man in his mid-thirties, married, a fairly good-looking guy but going to seed a bit (you'll see why). His closest partner in life isn't his wife but his daughter, age twelve. A likely enough child, bright and inquisitive, but caught in a war between father and mother. The mother is older, mid-forties. A bit of a blusterer, she sincerely loves her family and is pained by her husband's using their daughter as a chess piece in a power struggle she doesn't want but that her husband insists on perpetuating. And then, there's his drinking. Most evenings involve him and a bottle of wine, from which he becomes increasingly garrulous and belligerent. Then he powers down, the wine acting as a sleeping draught, and he's asleep by nine.

CLEANUP

It gets worse. From April to September, when a relative's swimming pool is open, this man hangs out there many afternoons into the evening, drinking heavily. Often, he takes his daughter along, driving home intoxicated, oblivious to the risks to his daughter, himself, and others. Sometimes, he's so drunk he has his daughter—a twelve-year-old, mind you—drive them home at midnight. He even denies wrongdoing to the department of social services, though the daughter acknowledges her father's high-risk behavior.

So far, the story consists of a series of behaviors—the father does this, the mother does that, the father fails to do this, and so on. But time was when the father himself was twelve, a time in his life when his parents' marriage, dying, turned into a nightmare of screaming verbal duels, scenes of the terrified child being put in the middle to choose the "good" parent, and some nights of sexual abuse by an uncle who took advantage of the chaos. The bass clef of the pain of being born into a fallen world was now exacerbated by the treble clef, the specific melody of pain points, piercing needles to reinforce the deeper ache of his loss of Eden and the vague not-yet of heaven.

As the needles penetrate, the flesh, already well-versed in its campaign to usurp God's place, makes proposals. The internal influence peddling might go something like this:

- You are abandoned.
- You are only a useable object.
- *But* you can learn to hide these realities behind a shield of anger and sarcasm.
- *And* you can use your body (that's what women want, anyway) to get love. So, stay attractive and dress the part.
- When pain overwhelms you, there are *lots* of anesthetics.
- You can also hide behind skillfully deployed fast-talk, using it to get your way and to deceive others into thinking you're confident.
- Rationalize, blame, deny, turn the tables—these methods are at least 95 percent effective.

SYNTHESIS

- Again, if pain overwhelms you, grab the anesthetics. Alcohol is everywhere, no matter what your age.
- Emotions, too, can be weaponized: you can manipulate with sadness, create distance with anger, isolate in fear.

These are the hidden longings, the reasonings, the decisions, the behaviors, the emotions that collect together, driving more and more of his life. As ever-new circumstances march over the horizon of life, he adjusts, deploying the same survival style in new ways like a general redeploying regiments to meet a flank attack. For example, once he feels that pain and frustration of exposure, he simply stops all communication with his wife, drawing a freezing circle around himself that ices every angle of approach.

The story reveals the scary power of the flesh—scary, because it has the force to foster a scenario where a father drives drunk with the daughter he ostensibly loves and even puts the inexperienced twelve-year-old behind the wheel of a car at midnight on a much-traveled road. Then, to refuse the help of an agency called in to correct and support, and, on top of that, subject his wife to a cold suspension of love, punishing her for a perceived betrayal even as he betrays his daughter time and gain. The urgent agenda of his flesh blinds him to an insanity: his way of surviving life's pain increases the chance of death, injury, and/or legal and marital problems. It guarantees that his daughter will have mental health struggles, not to mention moral and ethical confusion. Yet, he defends himself by saying that it's all "no big deal." That's because the true big deal is that he get the outcomes he demands to make his life work. His toxic solutions to the pain of life pollute the days of the very daughter he fathered twelve years before. Bleached coral.

Wallace Stegner's novel *Crossing to Safety*[1] portrays convincingly and chillingly what Paul means when he says, "The mind set on the flesh is death" (Rom 8:6). Stegner doesn't use these terms at all, he just shows them at work. The main character in the novel is Charity Lang, a burning force from a New England matriarchy that, for historical and cultural reasons, had long learned to shape the men in their lives into docile go-alongs, moldable and closely

---

1. Stegner, *Crossing to Safety*.

scripted. Charity is married to Sid, who is physically strong and intellectually bright but no match for Charity in her will to power. To Stegner's credit, he makes her a deeply human figure: she can be generous and compassionate, wanting the true good of the other so long as it doesn't trample on her plans. And does she plan! Running a tight ship, she issues agendas for everyone, especially the other three primary characters. Sid takes the brunt of her controlling forcefulness, her my-will-be-done grip on life and relationships. The novel is a long wearing down of Sid as he finds her will inescapable. The other two characters, Larry and Sally Morgan, are like a Greek chorus responding to, lamenting, and commenting on the tense jockeying between Charity and Sid.

Early in the story, Sid and Larry, newly hired professors in the University of Wisconsin's English department, discuss with their wives the tenuousness of their positions, both having been hired for one year with no guarantee beyond that. Larry has just sold a short story to *The Atlantic*. Sid, a drink in his hand, says, "To all of us. May we all survive the departmental axe."

Larry responds, "What are you talking about? . . . I'm cannon fodder, I'm a nine-month wonder. But if anybody's in, you are."

Sid: "Don't kid yourself. Rousselot [department head] was inquiring delicately just the other day what I'm working on. By the time they vote in April or May you'll have a bibliography as long as your arm, and I'll have my little undergraduate poems."

After some discussion, Charity bursts out impatiently: "Oh, bosh, Sid! Have some confidence in yourself! You're a splendid teacher, everybody says so. Go on being that. If they demand publications, write some. Just take it for granted you're going to be promoted, and they won't have the nerve not to."

Larry, trying to save the group's declining mood, raises a toast: "Let us be unignorable."

To which Charity almost yells, "Exactly! . . . You have to take your life by the throat and shake it."

The narrative goes on: "And we all laugh. We sidle away from Sid's anxiety and whatever it is between him and Charity."[2]

2. Stegner, *Crossing to Safety*, 49–50.

SYNTHESIS

What's going on between Sid and Charity is relational pain for Sid as Charity continues a process of recruiting him inexorably into her life script. She bends him toward it through an entire painful novel that shows how a relationship becomes an arrangement more about script following than true dialogue, productive conflict, and growing intimacy.

Another instance as Charity forges ahead in her plan to whomp up another perfect outcome: Once or twice when the weather lets up, the Langs take a stepladder out to the two acres that they have bought in the suburb called Frostwood, and climb on its slippery steps to test the view or the solar exposure. For Charity made up her mind when Sid took the Wisconsin job that they were not going to be like other instructors, kept for three years and then turned out to start all over again in some other, probably lesser place. Sid was going to be so superior, and together they were going to make themselves so indispensable to the university and the community that there could be no question of their being let go. She spends the first year finding land she likes. They will spend this one planning the house they will build on it. No cautionary words have any effect on her. If you want something, you plan for it, work for it, and make it happen.

But Sid isn't so sure. He has come from money and doesn't want—by forging ahead to build the house—to appear to seek a position based on financial advantage. Here's how the argument unfolds, Sid speaking first:

> "I'll look arrogant. It'll look as if we thought we could buy our way to promotion, or as if we thought ourselves so grand we could assume it. There's absolutely no guarantee we'll be here longer than this year and next. Do you want to build it just to move out of it?"
> "Pooh," Charity said, "I'd like to see them uproot us. Just have some confidence."
> "Caution would be more appropriate."
> "No sir," she said. "You won't budge me."[3]

---

3. Stegner, *Crossing to Safety*, 86.

CLEANUP

Charity, true to her flesh's vow to grab destiny by the throat and shake it, declares herself the master of the situation. She clenches these outcomes in fierce fists and slowly grinds Sid into a despairing blend of frustrated compliance and weak protest. Sid, whose heart longs to immerse itself in reading and writing poetry while making a living, runs afoul of Charity's ambitions for him to become a much-published, tenured professor.

Years later, in one of the most affecting scenes in the novel, Larry and Sally have driven cross-country to Vermont. The aim is to reconnect with Sid and Charity at a generations-old family compound where the foursome had enjoyed many languid, lovely summers. The reconnecting is urgent, because Charity is dying of cancer. After arriving, Larry, left to himself awhile, stops by Sid's old workshop in which there is a small study, a sort of hideaway. A shelf of books. Larry ticks off the usual ones, things like *The Oxford Universal Dictionary*, *Roget's Thesaurus*. Pause. He notices an oddity: "One book was backwards in the shelf, its spine turned inward. When I put it straight I saw that it was a rhyming dictionary. Imagining him jamming it hurriedly out of sight when he heard footsteps. I was ashamed for him. After a minute I turned it back the way he had left it."[4] The scene bids us imagine: Sid thinks he hears Charity coming and hides his heart, turning the spine of the book inward just as he has retracted his own into hiding at strategic Waterloos throughout the novel.

It's a scene that calls for pondering. One human being's determination to make life work in specified ways has destroyed the strength (the spine) of another. While she loves her husband, Charity loves something else far more: the collection of outcomes she needs to experience success, which is her version of safety and control. This is her first love, and it excels all other loves in her life. Her idolatrous pursuit of them means she has to sacrifice on her idols' altars the strength of her husband. Think about it: as the idols "whisper and mutter" (see Isa 8:19) to her, they ask for a sacrifice, saying, "Are you willing to burn your husband's strength on our altars?" By the end of the book, we know that her answer is an unqualified yes.

---

4 Stegner, *Crossing to Safety*, 177.

SYNTHESIS

Of course, there's another side to this, which is Sid's willingness to have his strength torn away from him and burned. It's not as though he doesn't put up a fight at times. But all too often, he only pushes back against Charity's tactics, failing both to address and to help her address her deeper strategy. Of the two concepts—tactics and strategy—the latter is the larger concept. For example, a business may decide on a strategy of, say, within five years making an initial public offering of stock. One of its tactics in reaching that strategy may be to acquire one or more smaller businesses in an effort to boost profitability. Charity's tactics include bullying, commanding, failing to listen, mocking, willfulness, resistance, sarcasm, pressure, denouncement. Her strategy aims at making Sid into a respected member of elite society. Who is that for? Not for Sid but for herself. Her tactics wear Sid down, and he never looks beyond them to wonder (and to help her wonder) what is so important—what is at stake—for her in remaking him in the image of the elites.

The heartbreaking power of the flesh! It's willing to put a twelve-year-old at the wheel of a car that's a hair away from becoming a death trap or a weapon or both. It can also grind a husband into a faint shadow of the man God made him to be while leaving his wife in the throes of a monstrous compulsivity so blind she can't discern that she's destroying a loved one who bears God's own image. Even as she racks up the outcomes on which she insists, she carves away the best of the man she purports to love. And it's not that she doesn't love him. Again, she just loves something else far more: wringing her destiny out of life at all costs. She sets up a win-lose scenario where she "wins" and the real Sid dies. By the same token, the inebriated father obtains the feel-good outcomes he demands, numbing himself with alcohol and a "party atmosphere." That's a "win" for him. But it "requires" that his daughter drive the potentially weaponized car. Of course, his daughter, even if she avoids wrecking the car, already suffers the wreckage of being unprotected, confused, and anxious. "The mind set on the flesh is death" (Rom 8:6).

These tragic situations, and many others like them, are the reason I've written this book. People are not fixed entities who can't help what they do. Rather, people do what really matters to them, and if they can change their sense of what matters, there's hope they

CLEANUP

can flex and change. In a word, repent. A fellow Christian and I were talking about a painful marriage we'd both observed for years. My friend said, "Everybody has crap they have to deal with." The gist of what he meant was "The wife in that marriage will continue to act self-servingly, and so will the husband. They will continue to react to each other, so you just take the good with the bad and move on." Stuff happens—just accept it. Don't drive yourself nuts wishing things were different.

Yet, is this the best ethic we have as Christians? That is, that we're stuck in behavioral loops and unbreakable patterns until we die? The tetralogy of which this book is the second part says no! In Christ, we're called to a better story than being stuck and playing out the sad string until the grave swallows us. In other words, we need not simply wait to die, self-shackled in a repetitive stage play of our own design. Jesus comes to burn the theater down. Discipleship consists of cooperating with Jesus the arsonist. Changing the image, Christ himself says, "Do not think that I came to bring peace on the earth; I did not come to bring peace, but a sword" (Matt 10:34). While the context of the verse points to social relationships, should we not also take Jesus to mean that he intends to take a sword to our own flesh? Paul seems to think so: "If by the Spirit you are putting to death the deeds of the body, you will live" (Rom 8:13). Here, "body" is equivalent to "flesh," as the context makes clear. As Jesus brings the sword against our indwelling sin, Paul teaches that we're to cooperate with him in wielding it. And this means war within. Nor should we be surprised at this when, elsewhere, Paul says, "For the flesh sets its desire against the Spirit, and the Spirit against the flesh; for these are in opposition to one another" (Gal 5:17). Of our indwelling sin (flesh), John Owen says, "There is no safety against it but in a constant warfare."[5]

Our ethic as Christians, then, is not that of loyalty to habitual patterns of self-interest until we die. Rather, we're to live in combat as warriors who, helped by the Holy Spirit, make every effort to deprive the flesh of vigor. This means that the norm of the Christian life is not that of crusted-over habits in what Abraham Heschel calls

---

5. Owen, *Temptation and Sin*, 12.

SYNTHESIS

"the wilderness of careless living,"[6] for to "hoard the self is to grow a colossal sense for the futility of living."[7] Rather than such hoarding, we're to break through the crust through self-examination, resisting the pull of the flesh to perpetuate our cherished outcomes. In Stegner's novel, Charity steams forward like an icebreaker, seeking to reach the promised land of the outcomes she has guaranteed herself. In the process, she has to suppress any knowledge of her actual impact on Sid. As Heschel observes, "The decay of conscience fills the air with a pungent smell."[8] Much of the novel pivots on whiffs of that stench, but no one locates it. Charity's death from cancer implies a metastasis of a cankered soul that remains elusive. Unable to see the wonder that is another person, she sees the world as marketplace and a well-trained Sid as currency she'll use for bartering.

The only route out of such tragic immolation is repentance. In one of Flannery O'Connor's short stories, she sets up a competitive tension between a grandfather and grandson. The struggle between them intensifies in their trip to the distant big city. In various ways, each one-ups the other until things reach such a pitch that the grandfather resorts to hiding when the boy falls asleep. When the grandson wakes, grandfather watches from concealment as the panicky boy trembles and moans, thinking himself lost in all that vastness. When the boy bolts off wildly, he crashes into an elderly woman, and they sprawl into the street together in a tangle of shock and misery. The grandfather comes puffing up, but when his grandson reaches out in urgent relief, the old man says, "This is not my boy. I never seen him before."[9] The disavowal sends them both reeling. The women around them reel, too, and recoil. Together again and silent, the two end up truly lost. Finally, the grandfather and grandson grope their way out of the city and find the homebound train. The grandfather cannot believe the sin he has committed. As he realizes they are going to make it home, he experiences the shock of recognizing grace, as we saw in chapter 3. O'Connor's brilliant

---

6. Heschel, *God in Search*, 141.
7. Heschel, *Man Is Not Alone*, 194.
8. Heschel, *Man Is Not Alone*, 152.
9. O'Connor, "The Artificial Nigger," in *Complete Stories*, 265.

depiction of God's mercy as it penetrates and softens a tough, worldly-wise codger invites us to the miracle of repentance and newness. Instead of damaging his grandson in an ever-worsening spiral of competition, a trade-war of hatred, he can begin to make amends, loving his grandson and building him up. He can say "you" to his grandson instead of "it" and encourage him to do the same. Mr. Head, the grandfather, receives a liberating mercy, all unlooked-for. He wakes up. He softens. He repents.

But what if one awakens and the other doesn't? What if "the action of mercy"[10] inspires the grandfather while the boy shrugs it off? What if love goes unanswered? What if one comes "to His own, and those who were His own did not receive Him" (John 1:11)? Then unmet love must hope, pray, and wait for new invitations and repentances in both oneself and the beloved. Then we must turn to the issues of time and of waiting, and that takes us to the third book in this tetralogy.

---

10. O'Connor, "The Artificial Nigger," in *Complete Stories*, 269.

Bibliography

Abbott-Smith, George. *A Manual Greek Lexicon of the New Testament*. Edinburgh: T. & T. Clark, 1937.

Argyris, Chris. *Overcoming Organizational Defenses: Facilitating Organizational Learning*. Upper Saddle River, NJ: Prentice Hall, 1990.

———. *Strategy, Change and Defensive Routines*. Boston: Pitman, 1985.

Augustine. *Confessions*. Translated by Henry Chadwick. New York: Oxford University Press, 1998.

Bachelard, Gaston. *The Poetics of Space*. Translated by Maria Jolas. Boston: Beacon, 1969.

Brueggemann, Walter. *Interpretation and Obedience*. Minneapolis: Fortress, 1991.

———. *The Message of the Psalms*. Minneapolis: Augsburg, 1984.

———. *Old Testament Theology: Essays on Structure, Theme, and Text*. Edited by Patrick D. Miller. Minneapolis: Fortress, 1992.

Brunner, Emil. *Man in Revolt*. Translated by Olive Wyon. Philadelphia: Westminster, 1947.

Buber, Martin. *I and Thou*. Translated by Walter Kaufmann. New York: Touchstone, 1996.

Crabb, Larry. *Shattered Dreams: God's Unexpected Pathway to Joy*. Colorado Springs, CO: Waterbrook, 2001.

Eliot, T. S. "Burnt Norton." In *Four Quartets*, 13–20. New York: Harcourt Brace, 1980.

Harrison, Everett F. "Worship." In *Baker's Dictionary of Theology*, edited by Everett F. Harrison, 560–61. Grand Rapids: Baker, 1960.

Heschel, Abraham. *God in Search of Man: A Philosophy of Judaism*. New York: Farrar, Straus & Giroux, 1983.

———. *Man Is Not Alone: A Philosophy of Religion*. New York: Farrar, Straus & Giroux, 1979.

Hopkins, Gerard Manley. *A Hopkins Reader*. Edited by John Pick. New York: Image, 1966.

John of the Cross. *The Dark Night of the Soul*. Translated by Edgar Allison Peers. New York: Doubleday, 1990.

BIBLIOGRAPHY

Jones, Alan. *Soulmaking*. San Francisco: HarperCollins, 1989.

Jung, Carl. *Psychology and Religion: West and East*. Vol. 11 of *The Collected Works of C. G. Jung*. 2nd ed. Translated by R. F. C. Hull. Edited by Herbert Read et al. Princeton: Princeton University Press, 1973. https://www.jungiananalysts.org.uk/wp-content/uploads/2018/07/C.-G.-Jung-Collected-Works-Volume-11_-Psychology-and-Religion_-West-and-East.pdf.

Käsemann, Ernst. *On Being a Disciple of the Crucified Nazarene*. Grand Rapids: Eerdmans, 2010.

Kidner, Derek. *A Time to Mourn and a Time to Dance: The Message of Ecclesiastes*. Leicester: InterVarsity, 1976.

LaCocque, André. *Onslaught against Innocence*. Eugene, OR: Cascade, 2008.

———. *The Trial of Innocence*. Eugene, OR: Cascade, 2006.

LaCocque, André, and Paul Ricoeur. *Thinking Biblically*. Translated by David Pellauer. Chicago: University of Chicago Press, 1998.

Lane, Belden. *Backpacking with the Saints*. Oxford: Oxford University Press, 2015.

Lewis, C. S. *The Great Divorce*. San Francisco: HarperSanFrancisco, 1973.

Lovelace, Richard. *Dynamics of Spiritual Life*. Downers Grove, IL: InterVarsity, 1979.

MacLean, Norman. *Young Men and Fire*. Chicago: University of Chicago Press, 1992.

Merriam-Webster, s. v. "epilogue." https://www.merriam-webster.com/dictionary/epilogue.

Moltmann, Jurgen. *The Future of Creation*. Philadelphia: Fortress, 1979.

O'Connor, Flannery. *The Complete Stories*. New York: Farrar, Straus & Giroux, 1990.

Owen, John. *Temptation and Sin*. Vol. 6 of *The Works of John Owen*. Carlisle, PA: Banner of Truth, 1967.

Payne, Leanne. *The Healing Presence*. Westchester, IL: Crossway, 1989.

Pierce, Nora. *The Insufficiency of Maps*. New York: Atria, 2007.

Ricoeur, Paul. *Oneself as Another*. Translated by Kathleen Blamey. Chicago: University of Chicago Press, 1992.

Spurgeon, Charles. *The Treasury of the Old Testament*. Vol. 1. London: Marshall, Morgan, & Scott, n.d.

Stegner, Wallace. *Crossing to Safety*. New York: Random House, 1987.

Tolbert, Elizabeth. "Voice, Metaphysics, and Community." In *Pain and Its Transformations*, edited by Sarah Coakley and Kay Kaufman Shelemay, 147–65. Cambridge, MA: Harvard University Press, 2007.

Tolkien, J. R. R. *The Fellowship of the Ring*. Boston: Houghton Mifflin, 1965.

Willard, Dallas. *Renovation of the Heart*. Colorado Springs, CO: NavPress, 2002.

Wirzba, Norman. *The Paradise of God: Renewing Religion in an Ecological Age*. Oxford: Oxford University Press, 2003.

"The Wizard of Oz." IMDb. https://www.imdb.com/title/tt0032138/quotes/qt5706421.

www.ingramcontent.com/pod-product-compliance
Lightning Source LLC
Chambersburg PA
CBHW070512090426
42735CB00012B/2744